ANIMAL INTELLIGENCE

OXFORD PSYCHOLOGY SERIES

EDITORS

DONALD E. BROADBENT
JAMES L. MCGAUGH
NICHOLAS J. MACKINTOSH
MICHAEL I. POSNER
ENDEL TULVING
LAWRENCE WEISKRANTZ

1. *The neuropsychology of anxiety: an enquiry into the functions of the septo-hippocampal system*
   Jeffrey A. Gray
2. *Elements of episodic memory*
   Endel Tulving
3. *Conditioning and associative learning*
   N. J. Mackintosh
4. *Visual masking: an integrative approach*
   Bruno G. Breitmeyer
5. *The musical mind: the cognitive psychology of music*
   John Sloboda
6. *Elements of psychophysical theory*
   Jean-Claude Falmagne
7. *Animal intelligence* (*Proceedings of a Royal Society Discussion Meeting*)
   Edited by L. Weiskrantz

ANIMAL INTELLIGENCE

PROCEEDINGS OF
A ROYAL SOCIETY DISCUSSION MEETING
HELD ON 6 AND 7 JUNE 1984

ORGANIZED AND EDITED BY
L. WEISKRANTZ, F.R.S.

OXFORD PSYCHOLOGY SERIES NO. 7

CLARENDON PRESS · OXFORD
1985

Oxford University Press, Walton Street, Oxford OX2 6DP
London Glasgow New York Toronto
Delhi Bombay Calcutta Madras Karachi
Kuala Lumpur Singapore Hong Kong Tokyo
Nairobi Dar es Salaam Cape Town
Melbourne Auckland
and associates in
Beirut Berlin Ibadan Mexico City Nicosia

Oxford is a trade mark of Oxford University Press

Published in the United States
by Oxford University Press, New York

British Library Cataloguing in Publication Data
Animal intelligence.
1. Animal intelligence—(Oxford psychology series; no. 7)
I. Weiskrantz, Lawrence II. Royal Society
156′.39 QL785

ISBN 019 852 124 3

First published in *Philosophical Transactions of the Royal Society of London*, series B, volume 308 (no. 1135), pages 1–216

Printed in Great Britain for the Royal Society
at the University Press, Cambridge

PREFACE

The time seemed ripe, when the idea of a Royal Society Discussion Meeting was suggested in 1983, for a conferring of minds on 'animal intelligence'. A fruitful meeting was held in June 1984, leading to the publication of this volume, the contents of which also appeared in the *Philosophical Transactions of the Royal Society*, B **308**, 1–216, 1985.

The topic is one that has very long held a deep interest for a variety of disciplines and for the informed layman, but in recent years one can discern a convergence of somewhat disparate approaches using freshly developed techniques within enlarged perspectives. Thus, the seminal research on sign language in chimpanzees has raised many fundamental and controversial issues of scientific methodology and the nature of language, communication, and thought in animals. At a somewhat different but related level are the contributions of ethologists dealing with questions of biological constraints on learning, illuminated by sophisticated and sensitive observations together with controlled experimentation in the field. From within experimental psychology laboratories have come, among other developments, new approaches to the difference between relatively inflexible 'habits' and, in contrast, appreciation by animals of consequences of changes in sequences of events. Also, quite new evidence has been provided for a degree of perceptual categorization in animals that exceeds most previous expectations. Some of these developments have also appeared independently in the context of distinctions between dissociable capacities from neuropsychological studies of brain function in man.

All of these approaches admit of a set of contrasting capacities which, while not necessarily continuous with all of those of man, suggest types of processes shared probably by all vertebrates, constrained or enhanced by the particular demands of the animal's natural environment and its sensory and motor apparatus. Whether the appearance or power of any subset of intellectual capacities also reflects the brain:body-mass ratio or other aspect of encephalization, while probable, remains a matter for debate. What is no longer a matter for debate is that the principal route to all such questions lies in Fabre's dictum that 'to bring out truth, we must resort to experiment, which alone is able to some extent to fathom the obscure problem of animal intelligence...observation sets the problem; experiment solves it'.

It seems fitting that the Royal Society should have served as the setting for the meeting in which these various converging strands could interact: current developments and controversies can hardly be understood outside the context of Charles Darwin's concept of evolutionary continuity of mental function, which provoked intense interest in questions of animal intelligence by many of his contemporary members of the Society, such as George Romanes. While the present set of papers cannot be thought to be more than a passing record of the current state of play and a body of evidence in a gradually changing scene, subject itself to the forces of evolution, one hopes it will serve not only to stimulate those unfamiliar with the current state of development but also to remind those in the future of where the field was at this point in time. Had there been more time and space the survey could have filled in some of its omissions, for example, the sophisticated work being carried out by several workers on associative learning in invertebrates, but the range of contributions reported here cover a wide spectrum.

I wish to thank, on behalf of all the contributors to this volume, the officers and staff of the

Society for their efficient, generous, and gracious support unstintingly offered from the early stages of planning right through to completion of this volume. I am also grateful for the helpful discussions with many colleagues during the planning stages, especially to Dr R. E. Passingham. And finally, I am also very grateful to my wife for preparation of the index in record time.

October 1984 L. WEISKRANTZ

CONTENTS

CONTENTS

Phil. Trans. R. Soc. Lond. B **308**, 3–19 (1985) [3]
Printed in Great Britain

Introduction: categorization, cleverness and consciousness

By L. Weiskrantz, F.R.S.
Department of Experimental Psychology, South Parks Road, Oxford OX1 3UD, U.K.

[Pullout 1]

Various recurring themes in the history of the subject are reviewed. In the context of adaptation to a complex environment, one precondition for survival must be a capacity for object identity, which may be the most basic form of categorization. Evidence will be presented that suggests that the capacity is not learned. In considering learned associations among categorized items, a distinction is made between reflexive and reflective processes: that is between those associations in which a cue or signal provides an unambiguous route to the response, no matter how complex that route may be, in contrast to those in which learned information must be ordered and reordered 'in thought'. An example of one experimental approach to the latter is provided. Finally, the problem of conscious awareness is considered in terms of stored categorical knowledge and associations, on the one hand, and a system that monitors them, on the other. Neurological evidence of disconnections between these different levels is reviewed.

1. Historical contrasts over the past 100 years

Given the uncertainty, not to say controversy, that the term '*human* intelligence' provokes these days, there may well be a question about the profitability of discussing '*animal* intelligence'. 'What', asked Fabre, the great French naturalist, 'is human intelligence? In what respect does it differ from animal intelligence? What is instinct? Are these two mental aptitudes irreducible, or can they both be traced back to a common factor?...These questions are, and always will be, the despair of every cultivated mind, even though the inanity of our efforts to solve them urges us to cast them into the limbo of the unknowable' (Fabre 1919, p. 107). But, in spite of this, there is no doubt that for the past 100 years there has been almost an obsessional interest in discovering or uncovering the mental capacities of non-human creatures, or 'brutes' as they were called until about the turn of the century.

The floodgates were opened by many, but no doubt Darwin must take the main responsibility because the revolution he spawned made it necessary to consider the continuity between ourselves and other creatures. Given that we have mental capacities, it seemed reasonable not only to ascribe them to animals but to enquire into their relative development, and Darwin, while acknowledging that 'the difference between the mind of the lowest man and that of the highest animal is immense', was, nevertheless, convinced that the difference 'certainly is one of degree and not of kind. We have seen that the senses and intuitions, the various emotions and faculties, such as love, memory, attention, curiosity, imitation, reason, etc., of which man boasts, may be found in an incipient, or even sometimes in a well-developed condition, in the lower animals' (*Descent of Man* (Darwin 1871), pp. 125–6). Darwin with the passage of time appears to have become more confident of this congenial conclusion: in the *Origin of species*, some 14 years earlier (Darwin 1859), he allowed only that 'a little dose of judgment or reason often comes into play, even with animals low in the scale of nature' (p. 266). Earlier writers (and even some later ones, such as Jacques Loeb, who resisted the flood), could be mechanistic, in the literal sense, about animals, as was Descartes. John Locke (1690), while not quite relegating animals to the level of bare machines, nevertheless was 'positive...that the power of abstracting is not at all in them; [it is] an excellency which the faculties of brutes do by no means attain to' (1690, vol. 1, p. 126). Fabre, Charles Darwin's contemporary (1916, chapter vii) scornfully and mercilessly demolished a putative observation of 'reason' in the wasp by Erasmus Darwin and would not allow insects even the 'little dose of reason' that Charles Darwin was prepared to admit.

From Darwin derives not only the explicit assumption of animal–human continuity, but also the implicit assumption that behaviour patterns, from which intelligence must be inferred, are proper candidates for evolutionary selection, a tenet that became the foundation stone of ethology and was also seminal for many branches of psychobiology. In one of his notebooks Darwin wrote that 'he who understands baboon would do more towards metaphysics than Locke' (Notebook M, p. 84, 16 August 1838; cf. Gruber (1974) pp. 317–318). Darwin himself, however, is relatively silent, even impotent, when it comes to those behavioural descriptions that might lead to a deep or systematic understanding of the cognitive capacities of animals, let alone metaphysics, and his most concentrated treatment of reasoning, for example, in the *Descent of man* is little more than scattered anecdotes of doubtful value. His observational methods, so well adapted to the analysis of 'natural' and conspicuous forms of behavioural display in *The expression of emotions in man and animals*, fail lamentably when applied to the drawing of inferences about cognition. In writing about 'reflection', for example, he remarks that 'Australians, Malays, Hindoos, and Kafirs...frown, when they are puzzled. Dobritzhoffer remarks that the Guaranies of South America on like occasions knit their brows. From these considerations, we may conclude that frowning is not the expression of simple reflection, however profound, or of attention, however close, but of something difficult or displeasing encountered in a train of thought or action. Deep reflection can, however, seldom be long carried on without some difficulty, so that it will generally be accompanied by a frown' (1965 edition, p. 222). He goes on to describe how he generated such a state in some unwitting volunteers by 'making them believe that I only wished to test the power of their vision'. He proceeds to even more difficult realms – 'abstraction' and 'meditation': 'The vacant expression of the eyes is very peculiar, and at once shows when a man is completely lost in thought.' He recruited Professor Donders into measuring the divergence of the eyes in such a state, 'if the eyes be held vertically, with the plane of vision horizontal, amounting to 2° as a maximum'. His description then goes on very precisely concerning the greater degree of divergence, up to 3° 5′, as the head droops, and the eyes are turned a little upwards, and 6° or 7° if the eyes are turned still more upwards. He concludes the section by remarking that 'we can understand why the forehead should be pressed or rubbed, as deep thought tries the brain; but why the hand should be raised to the mouth or face is far from clear' (p. 228).

'It is something to observe', said Fabre (on whom Darwin admiringly bestowed the title of 'incomparable observer', but who was for his part an unremitting critic of Darwin, referring to him sardonically as 'O illustrious master'), 'but it is not enough: we must experiment, that is to say, we must ourselves intervene and create artificial conditions which oblige the animal to reveal to us what it would not tell if left to the normal course of events...To bring out the truth, we must resort to experiment, which alone is able to some extent to fathom the obscure problem of animal intelligence...Observation sets the problem; experiment solves it, always presuming it can be solved; or at least, if powerless to yield the full light of truth, it sheds a certain gleam over the edges of the impenetrable cloud' (1919, pp. 108–9). The Darwinian revolution made it permissible, even compulsory, to consider animal cognition, but without itself providing the methods for pursuing the matter. But others were not slow to try to find them. This cannot be the place for a review of this quite fascinating history, starting perhaps with Fabre himself, then George John Romanes, Darwin's friend and ardent admirer, the equally ardent Lloyd Morgan, and that remarkable intrepid experimentalist, L. T. Hobhouse who anticipated much of Köhler's and Yerkes' line of thinking, and also helped to put 'Mr' Thorndike in his place (Lloyd Morgan, who was well disposed to Thorndike's approach, always addressed him as 'Professor'), and H. S. Jennings. The history is very rich, and deserves a good review, but I will have time only to refer to illustrative bits as we turn to other matters.

In holding a meeting today on such a topic, we must try to see how the scene has changed over the past 100 years. One point that ought to be made straight away is that in many ways it has not changed: issues discussed today were discussed just as earnestly and perspicaciously 100 years ago. Thus, we have much flurry today about 'intentionality' as applied to animal behaviour (cf. Dennett 1983) but Jennings in 1905 made a very similar point, with infinitely more empirical experience of the subject than modern philosophers, when he remarks that 'we usually attribute consciousness...because this is useful; it enables us practically to appreciate, foresee, and control [an animal's] actions much more readily than we could otherwise do so'. The anti-intentionality writers also have their historical forebears: Pavlov wrote, 'in our "psychical" experiments on the salivary glands at first we honestly endeavoured to explain our results by fancying the subjective condition of the animal. But nothing came of it except unsuccessful controversies, and individual, personal, uncoordinated opinions' (1928, p. 50).

There is also the same debate between 'rules' and 'reference' in earlier writers' distinction between 'practical learning' and 'rationality' (for example, Lloyd Morgan 1900; Hobhouse 1901). Earlier writers were also not unconcerned with the question of use of tools, as in the cebus monkey of Romanes, whose sister observed it cracking nuts with a hammer (an observation that Hobhouse could not confirm himself with his own monkeys), and Darwin in *Descent of Man* cited a number of positive claims of tool use by apes, monkeys, and elephants. ('The tamed elephants in India are well known to break off branches of trees and use them to drive away the flies; and the same act has been observed in an elephant in a state of nature', p. 81.)

Needless to say, there was also the same debate about whether animals understood words in isolation, or had some linguistic comprehension. Darwin concluded that 'that which distinguishes man from the lower animals is not the understanding of articulate sounds, for, as every one knows, dogs understand many words and sentences. In this respect they are at the same state of development as infants, between the ages of 10 and 12 months, who understand many words and short sentences, but cannot yet utter a single word (*Descent of Man*, p. 85). But Lloyd Morgan writes wryly that 'when I said "whiskey" to my fox-terrier, he would at once sit up and beg; not because his tastes were as depraved as those of his master, but because the *isk* sound, common both to "whiskey" and "biscuit", was what had for his ears the suggestive value' (1900, p. 203). Later he remarks 'the animal "word", if we like so to term it, is an isolated brick: a dozen, or even a couple of hundred such bricks do not constitute a building, be it a palace or only a cottage; hen language, or monkey language is, at best, so far as we at present have evidence, an unfashioned heap of bricks'. In an earlier work (1890) he put the now familiar epistemological point that [whilst] 'the actions of the speechless child and our dumb companions show that they...are capable of forming mental products of the perceptual order,...we must not forget that we interpret the percepts of children and animals; that in so doing we cannot divest ourselves of the garment of our conceptual thought, that we cannot banish the Logos, and that, therefore, these percepts other than ours cannot be identical with ours, though they are of the same order, saving their conceptual element. We may put the matter thus:

$$\left.\begin{array}{l}(1)\ x\times\text{dog-mind}\\(2)\ x\times\text{cat-mind}\\(3)\ x\times\text{infant-mind}\end{array}\right\}=\left\{\begin{array}{l}\text{percepts to be interpreted in terms of (4), being}\\\text{analogous thereto but not identical therewith}\end{array}\right.$$

(4) $x\times$ adult human mind = the percepts of psychologists, named or nameable.

Nor were writers of the time unaware of the powerful dangers of imitation in yielding misleading 'Clever Hans' types of results: Hobhouse (1901) has a lengthy and detailed review of the importance of the presence of the experimenter in a variety of such claims. And given modern interest in sequential runs of behaviour, the earlier concern with 'counting' by birds, and the recognition of its importance, is worth noting. Interestingly, Romanes wrote at great length about the subject in a letter to *The Times* (19 September 1888), both about rooks and about his own observations on a chimpanzee (which, if true, makes a chaining explanation impossible). The similarity to interpretations of much more recent experiments on birds by, for example, Koehler (1951), is striking. This use of the medium of *The Times* for interpreting animal behaviour, incidentally, is by no means past: provoked by an account in that paper on 20 October 1981 of the study by Woodruff and Premack that reported, among other things, that the chimpanzee Sarah was 'the first non-human to be accused of lying', two days later no less a person than a past President of the British Academy, Sir Kenneth Dover, drawing on the authority of Robert Louis Stevenson, wrote in a letter to *The Times* that 'all intelligent dogs are accomplished and incorrigible liars'. A little before that (21 October 1971), the following letter from a titled lady appeared in that same paper.

> From articles and lectures by experts, I have always understood that dogs have no colour sense. The following incident may therefore be of interest to professional and amateur alike.
>
> Our Labrador dog (11 years old), has shown no interest whatsoever in black and white television. We have recently acquired a colour set, since when he sits gazing at the screen for lengthy periods. This afternoon he watched Mrs Margaret Thatcher addressing the Conservative Party Conference. She was wearing the conference badge (in colour red) and after watching for a short time, he went up to the screen and licked the exact spot where the red badge was showing.
>
> Our dog has not had the pleasure of meeting Mrs Thatcher and therefore it could not be a mark of affection. It would appear that he concluded that the badge was a piece of red meat being dangled before his eyes.

As it happens, Romanes' sister, 'a very conscientious and accurate observer', had an even more perspicacious dog (Romanes 1882, p. 455) who displayed 'a most unmistakable recognition of portraits as representatives of persons', the evidence for which is reported fully (but, alas, inconclusively). And, just to cite one more example, earlier writers were just as

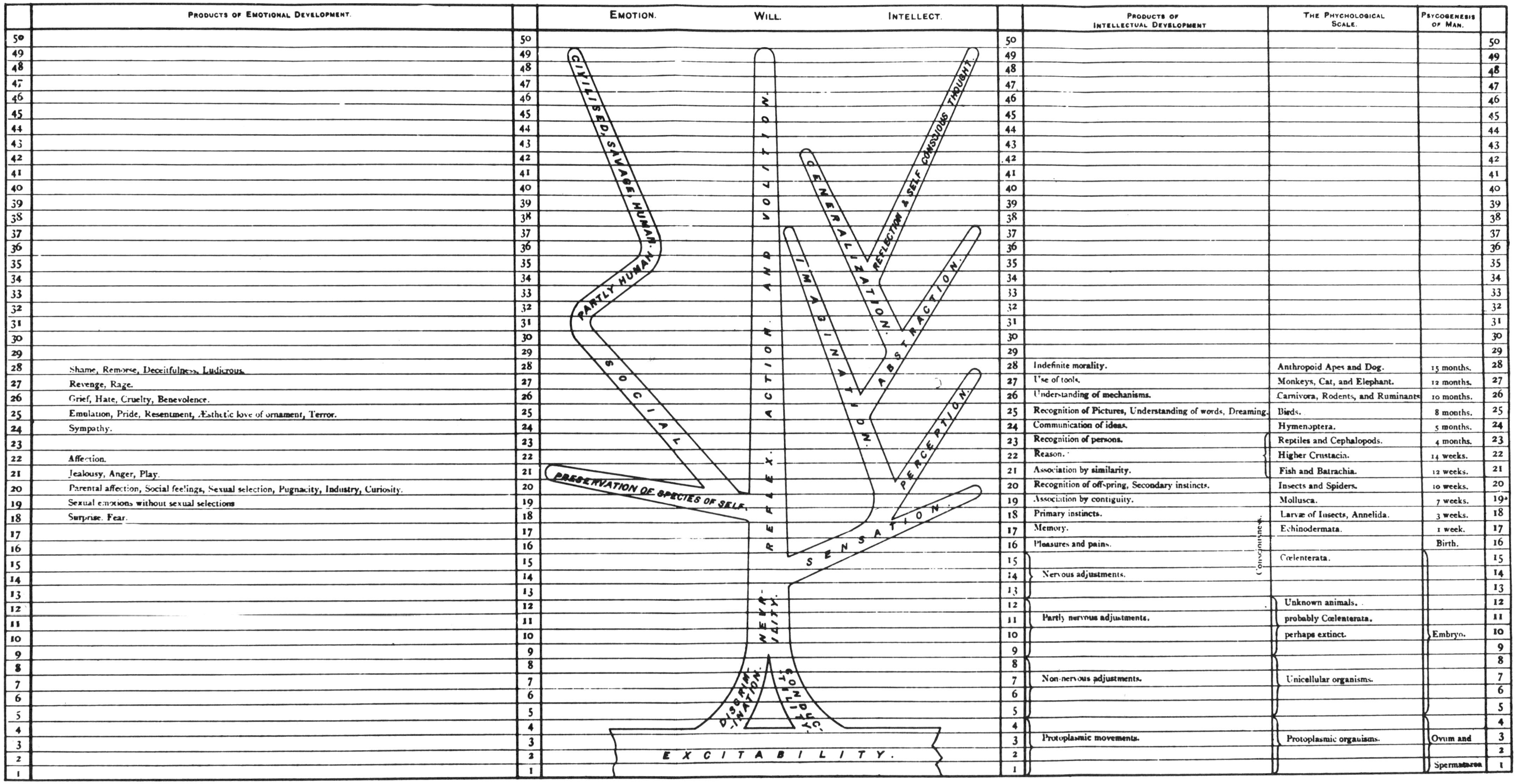

	Products of Emotional Development.		Emotion. Will. Intellect.		Products of Intellectual Development	The Psychological Scale.	Psychogenesis of Man.	
50		50		50				50
49		49		49				49
48		48		48				48
47		47		47				47
46		46		46				46
45		45		45				45
44		44		44				44
43		43		43				43
42		42		42				42
41		41		41				41
40		40		40				40
39		39		39				39
38		38		38				38
37		37		37				37
36		36		36				36
35		35		35				35
34		34		34				34
33		33		33				33
32		32		32				32
31		31		31				31
30		30		30				30
29		29		29				29
28	Shame, Remorse, Deceitfulness, Ludicrous.	28		28	Indefinite morality.	Anthropoid Apes and Dog.	15 months.	28
27	Revenge, Rage.	27		27	Use of tools.	Monkeys, Cat, and Elephant.	12 months.	27
26	Grief, Hate, Cruelty, Benevolence.	26		26	Understanding of mechanisms.	Carnivora, Rodents, and Ruminants	10 months.	26
25	Emulation, Pride, Resentment, Æsthetic love of ornament, Terror.	25		25	Recognition of Pictures, Understanding of words, Dreaming.	Birds.	8 months.	25
24	Sympathy.	24		24	Communication of ideas.	Hymenoptera.	5 months.	24
23		23		23	Recognition of persons.	Reptiles and Cephalopods.	4 months.	23
22	Affection.	22		22	Reason.	Higher Crustacia.	14 weeks.	22
21	Jealousy, Anger, Play.	21		21	Association by similarity.	Fish and Batrachia.	12 weeks.	21
20	Parental affection, Social feelings, Sexual selection, Pugnacity, Industry, Curiosity.	20		20	Recognition of offspring, Secondary instincts.	Insects and Spiders.	10 weeks.	20
19	Sexual emotions without sexual selections	19		19	Association by contiguity.	Mollusca.	7 weeks.	19
18	Surprise. Fear.	18		18	Primary instincts.	Larvæ of Insects, Annelida.	3 weeks.	18
17		17		17	Memory.	Echinodermata.	1 week.	17
16		16		16	Pleasures and pains.		Birth.	16
15		15		15		Cœlenterata.		15
14		14		14	Nervous adjustments.			14
13		13		13				13
12		12		12		Unknown animals.		12
11		11		11	Partly nervous adjustments.	probably Cœlenterata.		11
10		10		10		perhaps extinct.	Embryo.	10
9		9		9				9
8		8		8				8
7		7		7	Non-nervous adjustments.	Unicellular organisms.		7
6		6		6				6
5		5		5				5
4		4		4				4
3		3		3	Protoplasmic movements.	Protoplasmic organisms.	Ovum and	3
2		2		2				2
1		1		1			Spermatozoa	1

Consciousness.

Civilised, savage human.
Partly human.
Social.
Preservation of species or self.
Reflex action and volition.
Generalization.
Reflection & self consciousness thought.
Imagination.
Abstraction.
Perception.
Sensation.
Discrimination.
Conductility.
Excitability.

Figure 1. Scale of phylogenetic intelligence proposed by Romanes (1883).

(*Facing p.* 6)

explicit about both the possible purposes and origins of 'consciousness' as is, say, Sperry in his contemporary writings. Lloyd Morgan, whose strictures are often quoted incompletely, put the 'emergence' theory quite crisply (as well as a criticism of it), and despite his caution, introduced consciousness just as directly into the causal chain: 'consciousness is no longer merely a passenger in the ship of life. We may rather liken it to the captain of a modern ironclad, who, seated in the conning tower, directs all the movements and all the actions of the ship under his command' (1896, p. 276).

These are topics to which we shall return, but having pointed out similarities, we should turn to consider the differences between now and 1884.

First, there is no longer the concentrated concern with the comparative capacities of different species. This is not to say that there is not a strong interest, as we shall see from Professor Jerison's presentation, in correlations of cognitive capacities with encephalization, but this is a concern of the relation between cognition and the brain and its evolution, rather than with phylogenetic categories, as such. An example of a characteristically elaborate speculation of the 1880s is that of Romanes (figure 1, reproduced from original, 1883, including misspellings). But there was, even in Romanes' time, considerable disgreement about the relevance of phylogeny, although there was a splendidly direct attitude towards investigating it: Romanes (1882) freewheels through the animal kingdom in a single volume, reporting (mainly) anecdotal evidence on the emotional and intellectual powers of molluscs, ants, bees, termites, spiders, fish, reptiles, birds and mammals, with particular chapters on elephants, cats, foxes, dogs, monkey, apes and baboons. He enlisted the collaboration of his sister in making detailed observations on an allegedly tool-using monkey and the portrait-recognizing terrier. Hobhouse was much more direct: he simply went out and studied not only household pets, such as dogs and cats, but managed to do experiments himself on monkeys of several species, a chimpanzee, an elephant and an otter. He, along with Jennings (1904), would probably agree with Macphail's conclusions that there are no species differences that cannot be attributed indirectly to perception or motivation, but this point is bound to be controversial or at least tentative. Jennings took this very far: 'if *Amoeba* were a large animal, so as to come within the every-day experience of human beings, its behaviour would at once call forth the attribution to it of states of pleasure and pain, of hunger, desire, and the like, on precisely the same basis as we attribute these things to the dog' (p. 336). In the same vein, he later warns us that if *Amoeba* 'were as large as a whale, it is quite conceivable that occasions might arise when the attribution to it of the elemental states of consciousness might save the unsophisticated human being from destruction that would result from lack of such attribution' (p. 337).

Second, the main thrust of enquiry in the 50 years or so following *Origin of species* was on problem-solving. It was almost an account of how one would select animal candidates for the British civil service or the army officer corps. The extraordinary variety of problems set the great variety of animals by Hobhouse were ingenious and telling, as was Köhler's work on the Tenerife chimpanzees; they were a sort of ethology of animal cognition. Hobhouse had experiments with push-back bolts on boxes, stoppers, loops, weights, string-pulling, opening drawers, levers (including what could be called a Skinner box), door-pushing, sliding lids, pushing food out of long tubes with sticks, to mention just a few. Yerkes had an almost identical food-in-tube problem, also for chimpanzees, without acknowledgment of Hobhouse's work. Köhler's highly influential work was similarly concerned with giving animals problems to solve, and a certain amount of the work of the Tolman school, not to mention Maier's work on problem-solving in rats, falls into the same category. The current work, in contrast, is concerned rather more with trying to construct the *structure* of an animal's psychological world, as in the spatial memory work of Olton and associative learning of Gaffan, and, of course, most notably the chimpanzee work of Menzel, Savage-Rumbaugh and Rumbaugh, and Terrace. The question is not whether an animal is clever, but in which way does an animal remember and *think*.

Third, today's efforts are much more systematic and focused: perhaps not surprisingly, given the plethora of revelations over the past 100 years of how one can be misled. Rather than ranging widely over a number of species and a number of problems, today's investigators are much more focused on a particular aspect of one species' performance. Virtually every speaker at this meeting exemplifies this point, but Herrnstein's impressive work on categorization is a very good example. The ratio of empirical evidence to dogmatic assertion has increased gradually but very substantially, and there is sober recognition of just how much hard work is required to establish a fact. Fabre's dictum has gradually won out: '...a thousand theoretical views are not worth a single fact... Problems such as these, whether their scientific solution be possible or not, required an enormous mass of well-established data' (1919, p. 108). But this is, of course, relative. Hobhouse and Lloyd Morgan certainly focused much attention on the behaviour of the dog, and their descriptions, especially the former's, are exceedingly detailed and careful. Lloyd Morgan spent a great deal of time in observing the unsuccessful behaviour of a dog with a stick in its mouth when it comes to a narrow gap in a vertical grill or railing, and thought it an impossible task, even with training. 'Two of my friends criticized these results, and said that they only showed how stupid *my* dog was. *Their* dogs would have acted very differently. I suggested that the question could easily be put to the test of experiment. The behaviour of the dog was in each case – the one a very intelligent Yorkshire terrier, the other an English terrier – similar to that above described. The owner of the latter was somewhat annoyed, used forcible language, and told the dog that he could do it perfectly well if he tried' (1900, p. 143).

Fourth, the role of anecdotal evidence has changed, although one must be cautious about this: as Ghiselen remarked recently (1983), 'a respectable ethologist might call...anecdotal evidence nothing more than good observation in the field', and probably very few would agree with Seidenberg (1983) that 'behavior so novel that it can't be observed more than once can't be understood'. Many of today's anecdotes are contrived, they are experimental programmes in miniature, and even when not they are cast in the form of answering rather specific questions. This can be contrasted with the observation of Darwin's recounted in his letter to Romanes (20 August 1878, cited in Romanes 1896, p. 76), no doubt written with tongue pointing gently towards a cheek, but probably no more: 'Have you ever thought of keeping a young monkey, so as to observe its mind? At a house where we have been staying there were Sir A. and Lady Hobhouse [not, incidentally, *the* Hobhouse to whom we have been referring], not long returned from India, and she and he kept three young monkeys, and told me some curious particulars. One was that the monkey was very fond of looking through her eye-glass at objects, and moved the glass nearer and further so as to vary the focus. This struck me, as Frank's son, nearly two years old (and we think much of his intellect!) is very fond of looking through my pocket lens, and I have quite in vain endeavoured to teach him not to put the glass close down on the object, but he will always do so. Therefore I conclude that a child just under two years is inferior in intellect to a monkey.' Fabre, as usual, was scathing about such effortless comparisons: 'To disparage man and exalt animals in order to establish a point of contact, followed by a point of union, has been and still is the general tendency of the "advanced theories" in fashion in our day. Ah, how often are these "sublime theories", that morbid craze of our time, based upon "proofs" which, if subjected to the light of experiment, would lead to... ridiculous results' (1916, p. 128).

Fifth, there is a flowering of interest in cognition as nurtured by and revealed in social interaction. While Yerkes, as a relatively modern author, had some interest in chimpanzee diads, as of course do the Gardners and the Rumbaughs, few in the past have had the focused interest in social phenomena shown today by Hans Kummer, Jane Goodall, Robert Seyfarth and Dorothy Cheney, and Emil Menzel; and Nicholas Humphrey (1983) has argued that the demands of social knowledge are a major source of pressure for mental evolution.

Sixth, the fields of human cognitive psychology and cognitive neuropsychology have themselves developed in recent years in such a way that animal workers are asking new theoretical questions that stem directly from the human studies. Terrace's work on serial learning and knowledge of ordinal position is a nice example of such a development. Similarly, much current work on animal memory can be characterized in this way, drawing especially on evidence from human neuropsychology.

Seventh, there is the striking development of interest in human sign language as either taught to animals or as considered in relation to animal gestures. There was no lack of interest in the relevance of the deaf to animal communication in the 19th century. but the main impact was to relegate the deaf to the level of brutes. Thus, Romanes in an evening discourse to the British Association (16 August 1878) remarked: '...it occurred to me that a valuable test...was to be found in the mental condition of uneducated deaf-mutes. It often happens that deaf and dumb children of poor parents are so far neglected that they are never taught finger language, or any other system of signs, whereby to converse with their fellow creatures. The consequence, of course, is that these unfortunate children grow up in a state of intellectual isolation, which is almost as complete as that of any of the lower animals.' He goes on to say he has obtained all the evidence he could about such persons who happened to become educated afterwards. 'I find that their testimony is perfectly uniform. In the absence of language, the mind is able to think in the logic of feelings, but can never rise to any ideas of higher abstraction than those which the logic of feelings supplies. The uneducated deaf-mutes have the same notions of right and wrong, cause and effect, and so on, as we have already seen that animals and idiots possess. They always think in the most concrete forms, as shown by their telling us when educated that so long as they were uneducated they always thought in pictures. Moreover, that they cannot attain to ideas of even the lowest degree of abstraction, is shown by the fact that in no one instance have I been able to find evidence of a deaf-mute who, prior to education, had evolved for himself any form of supernaturalism.' He concludes, 'on the whole, then, from the mental condition of the uneducated deaf-mutes we learn the important lesson that, in the absence of language, the mind of man is almost on a level with the mind of a brute in respect of its power of forming abstract ideas'. What an extraordinary number of issues are compressed into this short extract!

Darwin was, of course, deeply interested not only in natural gestures and their role in communication, remarking that 'any one who has watched monkeys will not doubt that they perfectly understand each other's gestures and expression, and to a large extent, as Rengger asserts, those of man' (Darwin 1872, p. 60), but also in 'conventional signs which are not innate, such as those used by the deaf and dumb and by savages', with particular interest in the 'principle of opposition' within the sign language of, for example, the Cistercian monks (Darwin 1872, pp. 60–61). But the entire emphasis in Darwin is upon phylogenetic continuity: 'every true or inherited movement of expression seems to have had some natural and independent origin' (p. 355), and he would have welcomed Terrace's pertinent question as to 'how alarm calls of vervet monkeys differ from bird calls or from other familiar examples of animal communication. In each case there is a "vocabulary" of communicative acts' (Terrace 1983, p. 378). Darwin himself quotes an account that the *Cebus azaroe* in Paraguay utters at least six distinct sounds, and 'monkeys... when wild, utter signal-cries to their fellows; and... fowls give distinct warnings for danger on the ground, or in the sky from hawks (both, as well as a third cry, intelligible to dogs)' (Darwin 1871, p. 87). But Darwin's and his contemporaries' interest never extended, as far as I can discover, to the possibility of teaching such a system to 'brutes', which was the seminal contribution of the Gardners, nor to any deep effort to probe the linguistic properties of sign language or artificial 'token' languages.

Finally, today's work is set against a lengthy background of theories and systems of animal learning and memory that started with Pavlov and Thorndike, but has been richly, if contentiously, ramified in the work and theoretical structures of Tolman, Harlow, and many modern workers, as represented in our meeting by Mackintosh and Dickinson. Similarly, field work has flourished along with the fruitful developments in ethology, which have lent a sophistication to methods of observing animals plus a fertile background of knowledge concerning the daily habits of a variety of creatures in this field, as is evident in the contributions of Jane Goodall and Hans Kummer. And so this conference is one where these two major strands – careful field work plus the logical dissection of laboratory work – have an opportunity to converge.

2. Some illustrative current work in context

I want now to turn to some illustrative points that emerge from the work of some colleagues and myself, starting first of all with the capacity of abstraction which Locke, an empiricist (or, what Wolfgang Köhler used to call disparagingly, an 'empirist'), decreed was a capacity that brutes did not possess: it 'puts a perfect distinction betwixt man and brutes (Locke 1690, vol. 1, p. 126). Herrnstein (this symposium) and his coworkers, Morgan *et al.* (1976) and others, have shown that pigeons are exquisitely capable of grouping pictures of things, for example, trees, fish, oak leaves, faces, letters of the alphabet, symmetrical compared with asymmetrical shapes, efficiently and quickly into classes. I was struck by a comment of Herrnstein's in a recent paper (1984): 'To categorize, which is to detect recurrences in the environment despite variations in local stimulus energies, must be so enormous an evolutionary advantage that it may well be universal among living organisms. Seen in this light, categorization is just object constancy, which is perhaps the fundamental constancy toward which all other perceptual constancies converge. Rather than psycholinguistics, it is the psychology of perception and stimulus discrimination that impinge most directly on categorization.'

For some time I have been pursuing experimentally just this fundamental constancy in monkeys. The fundamental importance of object constancy for an animal's survival in the real world of stable objects is, I believe, hard to over-estimate, and yet we know very little about the mechanisms. The world is not a buzzing confusing place even, I suspect, to a newborn animal, but the sense organs are buzzing with a continuous flow of changing inputs.

In our work we presented the animal with a stable view of a real object, which was always rewarded with food, and after the animal learned to a criterion of 90% correct, we randomly injected probe trials in which the objects were altered in orientation, size, or texture-shadow arrangement (figures 2, 3 and 4). To teach the animal the original discrimination, the rewarded object must be contrasted with a negative object. We used new and different negative objects on every trial, so that we did not have to worry about making equivalent transformations for both negative and positive stimuli. The details are described in Weiskrantz & Saunders (1984).

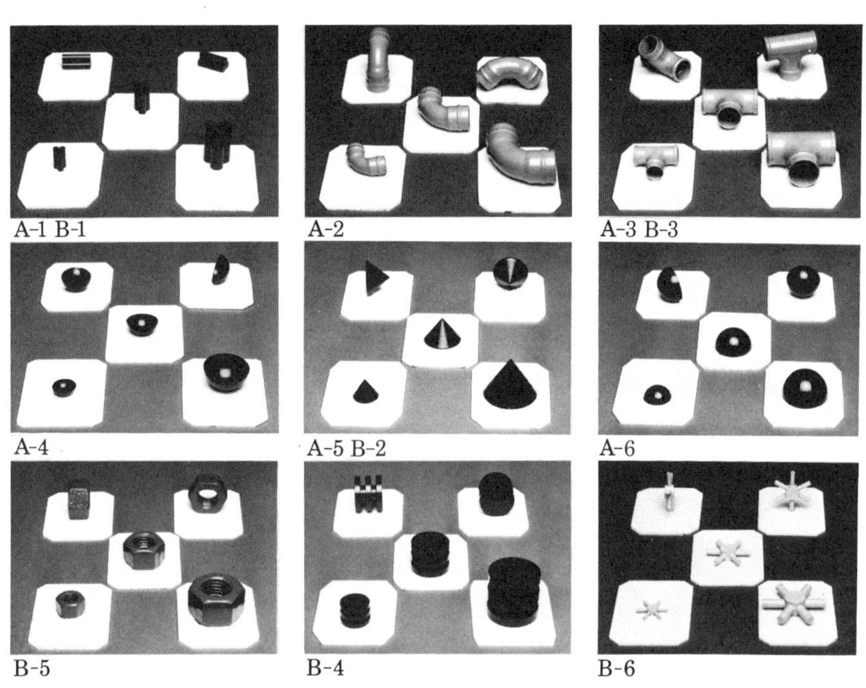

FIGURE 2. Photographs of stimulus objects with size and orientation transforms. View of object in the centre of each panel is of standard training object. The forward views are of the two size transforms, and the back views are of the two orientation transforms. Each white base plaque is 7.5 cm wide. (Reproduced by permission from Weiskrantz & Saunders (1984).)

On the very first problem in which naive animals were given their very first choice with transforms of the original object, they performed at well above chance on all three types of transforms, even though our training method would have biased them towards rejecting them (because the animals always failed to obtain reward for choosing novel objects). If we group together the first transform trials of the six problems used per animal in the study, a naive group performed at about 80% correct (chance being 33.33%). With more test sophistication, performance can become very efficient indeed. Our best group made only about 6%, 10% and 2% errors on the first occasions on which they saw shadow, orientation, and size transforms (out of 48 trials per transform type). It must be stressed that we can be sure that these transforms had never been seen before by the animals in their entire lives, and therefore they necessarily stimulated a fresh grouping of, for example, 'orientation' neurons in the striate cortex. And yet each virgin image was able to address the canonical or prototypical store with impressive efficiency. We have also tried to discover whether some regions of the brain are critical for this capacity, and it appears that the neocortex of the temporal lobe plays an especially important

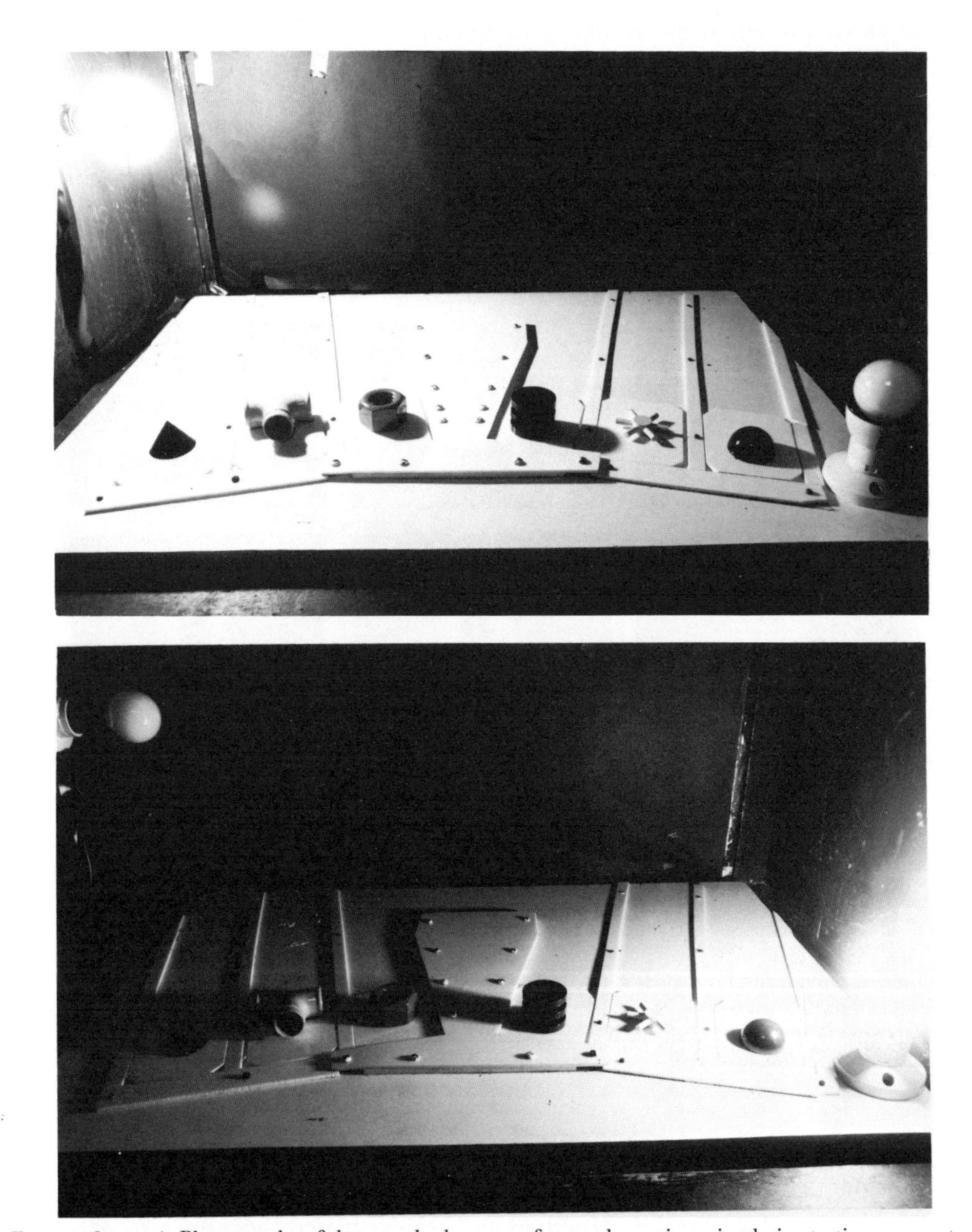

FIGURES 3 AND 4. Photographs of the two shadow transforms, shown in a six-choice testing apparatus. (Reproduced with permission from Weiskrantz & Saunders (1984).)

role (figure 5). Our rough scheme of the organization in the monkey is shown in figure 6. We also know from the work of Warrington and her coworkers (cf. Warrington 1982), which originally inspired our own study, that patients with right posterior hemisphere lesions can be severely impaired in identifying transforms ('unusual views') of familiar objects, without suffering from any difficulty in identifying the prototypical view or any loss in acuity or, apparently, any other sensory capacities.

No one knows how widespread the capacity is in the animal world: do crabs have it? Köhler (1915) made observations suggesting that young hens have it as regards shadow transforms. Marr (1982) has speculated on the computational aspects of the problem, but no one knows

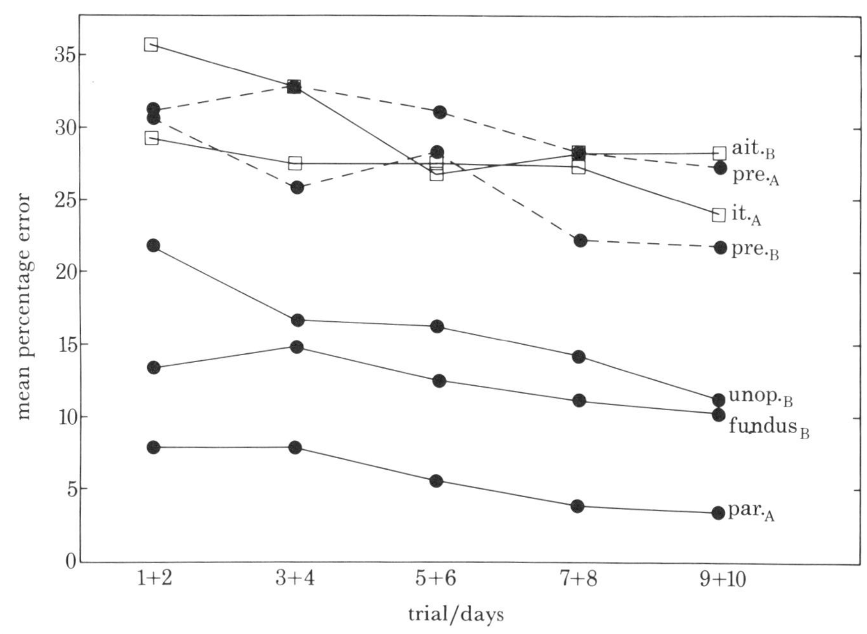

FIGURE 5. Mean percentage errors on transforms as a function of trials within all problems for each group of monkeys. Top three groups had lesions in various regions of inferotemporal neocortex: bottom three groups were controls. (Details in Weiskrantz & Saunders (1984).)

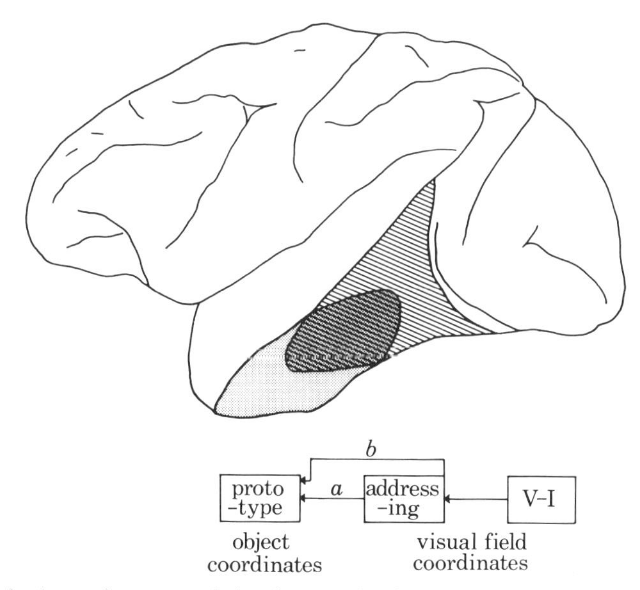

FIGURE 6. Hypothetical schema for stages of visual processing in striate, posterior and anterior inferotemporal cortical regions. (Reproduced with permission from Weiskrantz & Saunders (1984).)

in detail how the brain actually generates a stable representation of an object, with a coordinate system of its own, out of coordinates that are 'glued' to the animal's own retina. This is not the place to speculate, but merely to comment that it is an act of possibly universal but automatic intelligence. The achievement is clever, more clever than any machine has yet achieved, and is a prerequisite, perhaps even the foundation, for all visual abstraction and categorization, which is the attachment of identity of a thing with itself or of groups of things together. 'If a being has the power of thinking "thing" or "something", it has the power of transcending space and time.... Here is the point where intelligence ends and reason begins.' So Lloyd Morgan quotes Mivart with enthusiastic approval (1890, p. 375).

But such cleverness is unthinking, what Lloyd Morgan would call 'practical' intelligence. 'I wish [in no way] to disparage intelligence. Nine-tenths at least of the actions of average men are intelligent and not rational. Do we not all know hundreds of practical men who are in the highest degree intelligent, but in whom the rational, analytical faculty is but little developed? Is it any injustice to the brutes to contend that their inferences are of the same order as those of these excellent practical folk?' Trying to make amends to his brutes, he adds, 'if I deny them self-consciousness and reason, I grant to the higher animals intelligent inferences of wonderful accuracy and precision...in some cases, no doubt more perfect even than those of man, who is often distracted by many thoughts' (1890, p. 377). Hobhouse, similarly, concluded after the massive survey of his own experiments, 'none of my animals (with the possible exception now and again of the monkeys) showed the least understanding of the how or why of their actions, as distinct from the crude fact that to do such and such a thing produced the result they required.... What Jack (his dog) or the elephant knew was, crudely that they had to push [a] bolt.... The reason why...they obviously never grasped' (1901, p. 235).

Elsewhere Lloyd Morgan (1900, p. 59) makes a distinction that follows on from our own progression: he talks of a simpler level as belonging to the 'perceptual stage' and of a higher 'ideational' stage (following Stout) which embraces reflective thought, requiring 'deliberate attention to the relationships which hold good among the several elements of successive situations'.

Psychologists, from Köhler onwards, have in fact struggled with the question of reflective thought in animals, in just this sense of examining an animal's 'attention to the relationships which hold good among the several elements'. They tried, with some ingenuity and persistence, to put the matter under direct observation in controlled quasi-laboratory situations, within broadly defined theoretical frameworks, such as Tolman's. Perhaps the person who concentrated most strongly upon the problem was N. R. F. Maier (1929), who studied 'reasoning' in rats, by which he meant seeing whether an animal given experience of different spatial layouts could put them together to solve a new spatial problem. One of the difficulties in the work, which was heroic in scope, was in defining the individual 'elements', which were complex, and also, as in Köhler's work, the animal's past experience, which Harlow showed to be of such central importance in transforming slow learning into 'insightful' learning.

We have recently explored a different approach to the same problem, starting with observations from the neurological clinic. There are patients, about whom I shall say more later, who are unable to learn lists of paired-associate words. This is a task that is known to be heavily influenced by the capacity to link two supposedly unrelated words together by images in thought. Saunders and I have tried to see whether we could teach an animal a paired-associate task, not of words of course, but of visual objects. The strategy was to train the animal in stages on the following task, where each letter stands for a particular object and each + means food reward and each − means no food reward. (The details are in Saunders (1983).)

$$\overset{+}{A}\,\overset{+}{B} \text{ versus } \overset{-}{A}\,\overset{-}{C} \qquad \overset{-}{D}\,\overset{-}{B} \text{ versus } \overset{+}{D}\,\overset{+}{C}$$

$$\overset{+}{A}\,\overset{+}{B} \text{ versus } \overset{-}{D}\,\overset{-}{B} \qquad \overset{+}{D}\,\overset{+}{C} \text{ versus } \overset{-}{A}\,\overset{-}{C}$$

It will be noted that each object is equally often rewarded and non-rewarded, depending on its partner. If we examine the top two discriminations, it will be noted that B is positive and C is negative when paired with A, and just the reverse when each of these is paired with D. In the final stage, we presented one of the objects in isolation, and asked the animal, in effect, to 'think': what had it gone with when rewarded? If A, then B was the correct object, and not C. If D, then C and not B.

We found positive evidence that monkeys could do this, our control being the presentation of the last stage without experience of the previous pairings. Most interestingly, with lesions meant to simulate some of the same features as seen in patients, the crucial group of animals had no difficulty in learning the original task, but were impaired in the final stage. And so learning that an object or a pair of objects inevitably leads to food is not sufficient to enable the animal to reflect on which element went with which element.

Other varieties of stimulus–stimulus association tasks will be discussed by Gaffan. And Passingham and Dickinson, in different ways, both address the same type of issue relevant to this issue: what is the difference, theoretically and experimentally, between a task, on the one hand, that has become habitual, automatic, 'reflexive', and in which the information at hand leads directly to the acquired behavioural act (even if that task initially was complex and demanding, for example driving a car) and, on the other hand, a 'reflective' task in which the animal is able to make a fresh appraisal or reappraisal of its stored knowledge. The study of brain disorders appears to make some such distinction very important (cf. Weiskrantz 1982).

The neurological clinic also provides observations relevant to that most enduring of questions: consciousness. Like sin or poverty, the problem of consciousness will not go away, despite strenuous efforts to exorcise it. Many tried to argue that it is just an epiphenomenon, what T. H. Huxley (quoted by Lloyd Morgan) dismissed as 'simply a collateral product...as completely without any power...as the steam-whistle which accompanies the work of a locomotive engine is without influence on its machinery'. Even Lloyd Morgan could not stomach Huxley's 'steam-whistle' analogy, and replied: 'It is nothing less than pure assumption to say that the consciousness, which is admitted to be present, has practically no effect whatever upon the behaviour. And we must ask any evolutionist who accepts this conclusion, how he accounts on evolutionary grounds for the existence of a useless adjunct to neural processes' (1900, pp. 308). That the history of its conceptual status is tangled and fraught is beyond dispute, and the supposed criteria for recognizing it vary from 'learning' (Romanes), capacity for choice (Lloyd Morgan), to, among other things, 'the possession of ethical values' (Hubbard 1975). William James stressed purposive behaviour as the essential criterion: 'the pursuance of future ends and the choice of means for their attainment' (1890, p. 8), a definition that allowed him readily to endorse the view of some of his contemporary physiologists that the spinal cord of the frog displays 'conscious intelligence'. The criteria, it is clear, are as varied as the properties of adaptive behaviour themselves, and less amenable to enquiry. Here I wish to concentrate on one aspect of the problem: acknowledged awareness, the capacity to know what one is doing.

Perhaps we might acknowledge that much of the time we are unaware of our moment-to-moment activities: we can drive a car, walk down the street, wave our hands as we speak, or indeed even hold conversations about the weather, in states that are little better than sleep-walking. Of some of our responses, for example, pupillary adjustment, we are never aware. It is fortunate that we do not waste time on reflecting on such activities, or else there would be little capacity left for those activities, simultaneously present, that really do benefit from our reflection. The striking feature from neurology is that damage to certain regions of the brain prevents patients from having knowledge even of those processes for which we normally do acknowledge an awareness.

For example, patients have been reported with damage to the striate cortex who, at a very high level of proficiency, can locate visual targets in space, detect their onset, discriminate between gratings of different orientations, between moving and stationary stimuli, and yet do not acknowledge 'seeing'; they report they are merely guessing, a phenomenon that has been called 'blindsight' (Weiskrantz *et al.* 1974). The precise neurological status of 'blindsight' is a matter of controversy (cf. Campion *et al.* 1983) but not the reported phenomena themselves. The problem is not one merely of verbal disconnection, as in split-brain patients, because the discriminative response can be either verbal or non-verbal. And in this sense, the blindsight patients behave differently, in my experience, from those split-brain patients of Sperry's whom I have seen (but who also demonstrate, of course, another form of a clear dissociation between skilled discriminative capacity and acknowledgement of awareness (Sperry 1974). The problem for the blindsight patient is not whether he speaks or not but the type of question he is trying to answer and the type of decision he makes: it is the distinction between *reacting*, which he can do, and *monitoring*, which he is apparently unable to do, or is at least very deficient in doing.

Perhaps even more striking are those patients who can be shown to be capable of learning and retention, but who have no acknowledged 'memories' associated with doing so, the so-called amnesic syndrome patients. So ineffective are their own acknowledgements that for a long time the patients were thought actually to have lost the ability to learn anything new at all. But it is now clear from the work of many workers that these seriously disadvantaged patients can, for example, learn new motor skills, can show verbal and perceptual learning, can acquire conditioned reflexes and many other tasks, and yet show no recognition of the situations in which these tasks were presented or that their own performance has changed as a result of prior exposure. For example, a patient can acquire a conditioned eyelid response (figure 7), and yet be unaware of the (to us) highly memorable set of events. Nevertheless, he will respond appropriately to the light and tone that precede the puff of air to the eye, showing clear evidence of conditioning. When questioned (sitting in front of the apparatus) in the frequent brief rest pauses, he would simply invent accounts about what was going on: for example, that his knowledge of languages was being tested (Weiskrantz & Warrington 1979). We have argued (Warrington & Weiskrantz 1982) that the amnesic patient can learn and retain events that are reliably predictable, where there is redundancy and hence the possibility of automaticity either of forming habits, or of simply strengthening an item through sheer exposure, as in so-called 'priming' tasks. Inconsistency from occasion to occasion, as in a recognition situation with repeated stimuli, is apt to require reflection, matching, ordering, and reordering; what the amnesic patient is unable to do is manipulate his own memories 'in thought'. Indeed, for all of us it would seem that automatic memories are devoid of any awareness of being memories as such: we do not attach such a quality to our everyday words, although these obviously have

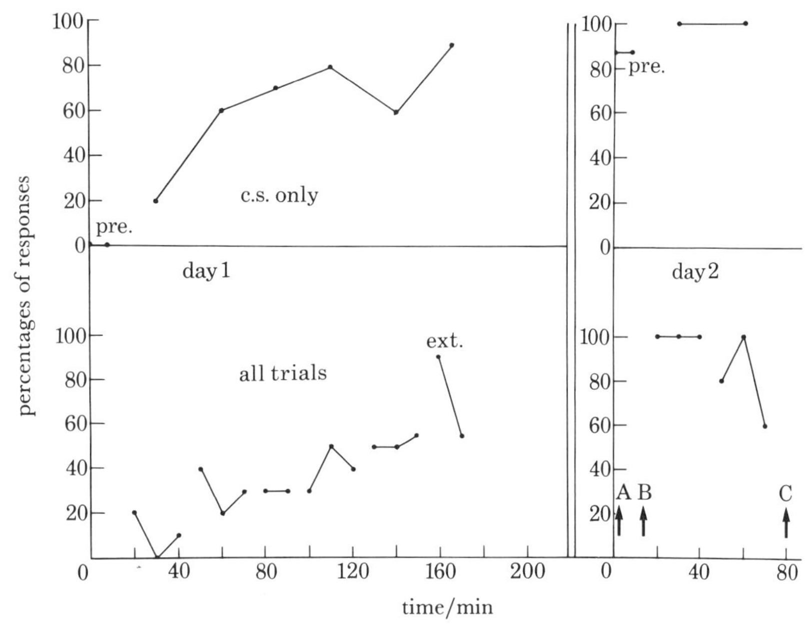

FIGURE 7. Percentage of conditioned eyelid responses for one amnesic subject. Top panel shows 'probe' trials in which the conditioned stimulus (c.s.) was delivered without the unconditioned stimulus on a random schedule. Bottom panel shows conditioning and all other trials combined. Interruptions in graph indicate occurrence of 10 min rest breaks in which subject was interviewed. (Reproduced with permission from Weiskrantz & Warrington (1979).)

been learned. We do not attach such a quality to seeing a traffic light: we stop without reflecting that we 'remember': red means stop. If we do reflect, we are apt to stop for more catastrophic reasons! The neurological cases are useful and striking in showing how complete can be the integrity of skills of perception and learning in the face of a dissociation between them and the patient's own knowledge.

We now know that both instrumental and classical conditioning can occur at a subcortical level in the absence of cerebral cortex in mammals, provided that the stimuli are not so highly patterned as to required cortical processing (Oakley 1981; Oakley & Russell 1977). The implication that I draw from this, together with the human neurological evidence, is that 'awareness' of our behaviour is a precious commodity reserved for rather special occasions when we wish to compare one visual image with another or one piece of stored information with another, that is, to imagine and to reflect, and so there is a link between our earlier concern with reflection and our present one with awareness. It would be a waste of cerebral processing to have to continue to dwell on redundant information or automatic control.

Now, if we grant such an argument, how would we know whether an animal is aware? It is by no means a trivial question, either philosophically or practically in these days of heightened sensitivity about animal suffering. Awareness in the animal kingdom has been treated as either equivalent to being responsive, which clearly will not do, as we have seen, or as a gradually emergent property of evolving systems. The former was perhaps Jennings' view, and was Lloyd Morgan's early view: 'Suppose...a young bird seizes a bee and proceeds to swallow it, but is stung in the process; ...every step of the process is taken in the field of his conscious experience...every bit of experience, no matter how trivial...exists as such for consciousness' (1896, p. 273). The emergence view has been championed by Sperry (1970), who is, however, not very explicit in telling us how we would recognize when emergence has been achieved. The approach, in practical terms, that seems to me to be the most direct and promising is that ingeniously initiated by Beninger *et al.* (1974), who gave rats the opportunity to indicate whether they discriminate their own acts. They allowed them to face-wash, or rear up, or walk, or remain immobile as they wished, but food reward was contingent on the animals pressing a different lever in association with each of these acts. Morgan & Nicholas (1979) have followed this up in an interesting study in which they argue that one of the reasons why it is difficult actually to train a rat to do certain things, like scratch, is that such acts cannot serve as 'an adequate discriminative stimulus', which in colloquial terms means that animals may not be aware of what they are doing. I suspect that it is equally difficult to train a dog to wag its tail for food reward.

The essential step in any such analysis lies in providing the animal or the person with a parallel 'commentary' response. Not only must he be able to discriminate between A and B by pressing one or another panel, say, but he needs a third key to comment upon what he is doing on the first two keys, to give a 'yes–no' commentary on whether the discrimination was 'guessing' or not, or better still, to give a confidence rating. In everyday discourse the commentary system for ourselves is verbal, but the principle is the same whether or not we actually use words. The evolutionary value of the third key capacity is not merely to enable raw sentience to occur, but to endow us with the capacity to *compare*, to put our intercourse with the world in the form of cognitively manipulable percepts and memories: in short, to enable us to think.

We do not know at what level of neural organization in evolution the third key would start to be effective, but the essential minimal organization to be sought would be one that could fulfil a monitoring function: it cannot be a mere 'emergent' property of complexity or cleverness as such. As Sperry has argued, 'it is not merely complexity or high order organization that...endows a neural event with conscious awareness. It is rather the specific operational design of the cerebral mechanism for the particular conscious function involved. The neural mechanisms for conscious experience are not just more complex, they are specifically structured on an operational, functional basis to create particular sensations, percepts, and feelings, and to provide a rapid representation of external reality' (1970, p. 589). We do not know now just how to recognize these neural mechanisms, but I believe, somewhat optimistically, that a Discussion Meeting held in 2084 will have something to contribute to the question. Meanwhile, our meeting in 1984 will no doubt illuminate lots of other issues in ways that would have fascinated and deeply impressed our forebears of 100 years ago.

References

Beninger, R. J., Kendall, S. B. & Vanderwolf, C. H. 1974 The ability of rats to discriminate their own behaviours. *Canad. J. Psychol.* **28**, 79–91.

Campion, J., Latto, R. & Smith, Y. M. 1983 Is blindsight an effect of scattered light, spared cortex, and near-threshold vision? *Behav. Brain Sci.* **6**, 423–486.

Darwin, C. 1859 *The origin of species.* London: Oxford University Press. (Page citations from 6th edn, 1951.)

Darwin, C. 1871 *The descent of Man.* London: John Murray. (Page citations from 1894 edn.)

Darwin, C. 1872 *The expression of the emotions in Man and animals.* Chicago: University of Chicago Press. (Page citations from 1965 edn.)

Dennett, D. C. 1983 Intentional systems in cognitive ethology: the 'Panglossian paradigm' defended. *Behav. Brain Sci.* **6**, 343–390.

Fabre, J. H. 1916 *The hunting wasps.* Translated by A. T. de Mattos. London: Hodder and Stoughton.

Fabre, J. H. 1919 *The mason wasps.* Translated by A. T. de Mattos. London: Hodder and Stoughton.
Ghiselin, M. T. 1983 Lloyd Morgan's canon in evolutionary context. *Behav. Brain Sci.* **6**, 362–363.
Gruber, H. E. 1974 *Darwin on Man.* London: Wildwood House.
Herrnstein, R. J. 1894 Objects, categories, and discriminative stimuli. In *Animal cognition* (ed. H. C. Roitblat, T. G. Bever and H. S. Terrace). Hillsdale, New Jersey: Erlbaum.
Hobhouse, L. T. 1901 *Mind in evolution.* London: Macmillan. (Page citations from 1926 edn.)
Hubbard, J. I. 1975 *Biological basis of mental activity.* Reading, Massachusetts: Addison-Wesley.
Humphrey, N. 1983 *Consciousness regained.* Oxford: Oxford University Press.
James, W. 1890 *The principles of psychology.* London: Macmillan. (Page citations from 1901 edn.)
Jennings, H. S. 1904 *Behavior of the lower organisms.* Bloomington: Indiana University Press. (Page citations from 1962 edn.)
Koehler, O. 1951 The ability of birds to 'count'. *Bull Anim. Behav.* **9**, 41–45.
Köhler, W. 1915 Optische Unterssuchungen am Schimpansen und am Haushuhn. *Abh. preus. Akad. Wiss.* (*phys.-math.*) **3**, 1–70.
Locke, J. 1690 *An essay concerning human understanding.* London: J. M. Dent & Sons, Everyman's Library. (Page citations from 1961 edn.)
Maier, N. R. F. 1929 Reasoning in white rats. *Comp. Psychol. Monogr.* **6**, no. 29, 1–93.
Marr, D. 1982 *Vision.* San Francisco: W. H. Freeman and Co.
Morgan, C. Lloyd 1890 *Animal life and intelligence.* London: Edward Arnold.
Morgan, C. Lloyd 1896 *Habit and instinct.* London: Edward Arnold.
Morgan, C. Lloyd 1900 *Animal behaviour.* London: Edward Arnold.
Morgan, M. J., Fitch, M. D., Holman, J. G. & Lea, S. E. G. 1976 Pigeons learn the concept of an 'A'. *Perception* **5**, 57–66.
Morgan, M. J. & Nicholas, D. J. 1979 Discrimination between reinforced action patterns in the rat. *Learn. Motiv.* **10**, 1–27.
Oakley, D. A. 1981 Performance of decorticated rats in a two-choice visual discrimination apparatus. *Behav. Brain Res.* **3**, 55–69.
Oakley, D. A. & Russell, I. S. 1977 Subcortical storage of Pavlovian conditioning in the rabbit. *Phys. Behav.* **18**, 931–937.
Pavlov, I. P. 1928 *Lectures on conditioned reflexes.* Translated by W. H. Gantt. New York: International.
Romanes, E. G. 1896 *The life and letters of George John Romanes.* London: Longmans, Green & Co.
Romanes, G. J. 1878 Evening discourse delivered before The British Association, Dublin. London: Taylor and Francis.
Romanes, G. J 1882 *Animal intelligence.* London: Kegan Paul, Trench & Co.
Romanes, G. J. 1883 *Mental evolution in animals.* London: Kegan Paul, Trench & Co.
Saunders, R. C. 1983 Some experiments on memory involving the fornix–mammillary system. D. Phil. thesis, University of Oxford.
Seidenberg, M. S. 1983 Steps toward an ethological science. *Behav. Brain Sci.* **6**, 377.
Sperry, R. W. 1970 An objective approach to subjective experience. *Psychol. Rev.* **77**, 585–590.
Sperry, R. W. 1974 Lateral specialization in the surgically separated hemispheres. In *The Neurosciences. Third Study Program* (ed. F. O. Schmitt & F. G. Worden), pp. 5–19. Cambridge, Massachusetts: M.I.T. Press.
Terrace, H. S. 1983 Nonhuman intentional systems. *Behav. Brain Sci.* **6**, 378–379.
Warrington, E. K. 1982 Neuropsychological studies of object recognition. *Phil. Trans. R. Soc. Lond.* B **298**, 15–33.
Warrington, E. K. & Weiskrantz, L. 1982 Amnesia: a disconnection syndrome? *Neuropsychologia* **20**, 233–248.
Weiskrantz, L. 1982 Comparative aspects of studies of amnesia. *Phil. Trans. R. Soc. Lond.* B **298**, 97–109.
Weiskrantz, L. & Saunders, R. C. 1984 Impairments of visual object transforms in monkeys. *Brain.* (In the press.)
Weiskrantz, L. & Warrington, E. K. 1979 Conditioning in amnesic patients. *Neuropsychologia* **17**, 187–194.
Weiskrantz, L., Warrington, E. K., Sanders, M. D. & Marshall, J. 1974 Visual capacity in the hemianopic field following a restricted occipital ablation. *Brain* **97**, 709–728.
Yerkes, R. M. 1943 *Chimpanzees.* New Haven: Yale University Press.

Phil. Trans. R. Soc. Lond. B **308**, 21–35 (1985) [21]
Printed in Great Britain

Animal intelligence as encephalization

By H. J. Jerison
University of California at Los Angeles, Department of Psychiatry, School of Medicine, Los Angeles, California 90024, *U.S.A.*

There is no consensus on the nature of animal intelligence despite a century of research, though recent work on cognitive capacities of dolphins and great apes seems to be on one right track. The most precise quantitative analyses have been of relative brain size, or structural encephalization, undertaken to find biological correlates of mind in animals. Encephalization and its evolution are remarkably orderly, and if the idea of intelligence were unknown it would have to be invented to explain encephalization. The scientific question is: what behaviour or dimensions of behaviour evolved when encephalization evolved? The answer: the relatively unusual behaviours that require increased neural information processing capacity, beyond that attributable to differences among species in body size. In this perspective, the different behaviours that depend on augmented processing capacity in different species are evidence of different intelligences (in the plural) that have evolved.

Structural encephalization is a morphological phenomenon, the enlargement of the brain beyond that expected from the enlargement of the body. Since brain size is proportional to neural information processing capacity (between species; see below), the evolution of encephalization was, essentially, the evolution of an increase in information processing capacity. This is obviously important for the evolution of animal intelligence, and it is one of the reasons for the interest in encephalization.

Among vertebrate species, brain size is determined mainly by body size. The relation is described as the brain–body allometric function, which has been estimated from the regression of log brain size against log body size in appropriate samples of species (Martin 1983). Encephalization is then the increase in brain size beyond that expected from the allometric brain–body relation. As a quantitative exercise, encephalization, or the increase in *relative* brain size, is often determined by calculating the residual for a species relative to the allometric regression; the residual is an 'encephalization quotient' (Jerison 1985).

The importance of relative brain size, or encephalization, has been recognized since classical times (see Jerison (1973, 1982) for historical reviews). In this essay I follow an outline implicit in a seminal statement by Karl Lashley, in his presidential address on the evolution of mind, presented before the American Society of Naturalists. He stated the issues as follows.

> The only neurological character for which a correlation with behavioral capacity is supported by significant evidence is the total mass of tissue, or rather, the index of cephalization...which seems to represent the amount of brain tissue in excess of that required for transmitting impulses to and from the integrative centres. (Lashley 1949, p. 33.)

I have a sentimental reason for organizing this essay according to Lashley's outline, because my first work in this area was inspired by these words. But I am also convinced that his approach leads to fundamental insights into the nature of intelligence as a biological phenomenon. 'Behavioural capacity' is clearly the same as 'animal intelligence', the central issue in this discussion. The index of cephalization is an encephalization quotient, which I will treat in a qualitative, pictorial way, rather than with numerical residuals. (To compute residuals we have to make assumptions that unnecessarily complicate the analysis and are not really needed here (see Jerison 1985).) The idea of 'excess brain tissue' takes us to the meaning of absolute brain size. This essay on intelligence, encephalization, and brain size may be thought of as a set of variations on Lashley's theme. I will change only the order of the topics, first discussing the evolution of encephalization and the meaning of brain size as background for the later discussion of animal intelligence.

Lashley began with the idea of behavioural capacity, or intelligence, as if there was consensus about its meaning. In the search for a neurological correlate for this 'well-understood' trait, he noted that at mid-century the only correlate that had been discovered was encephalization, but further research could presumably lead to the discovery of more and finer correlates. I propose that encephalization is, in fact, the fundamental trait and that it may be fruitless to seek finer correlates of intelligence. Finally, although there is no real consensus on intelligence, some unusual features of both human and animal intelligence about which we can agree may be better understood if we assume that the biological foundations of intelligence are in encephalization. I, therefore, begin with the neural correlate, and conclude by analysing animal intelligence in terms of encephalization.

1. Structural encephalization

Brain and body masses are known for hundreds of species of vertebrates and provide the fundamental data on structural encephalization. Minimum convex polygons drawn to contain the log data of each class provide the simplest and clearest picture of how the allometric and encephalization factors determine brain size. Figure 1 summarizes such data in the major living

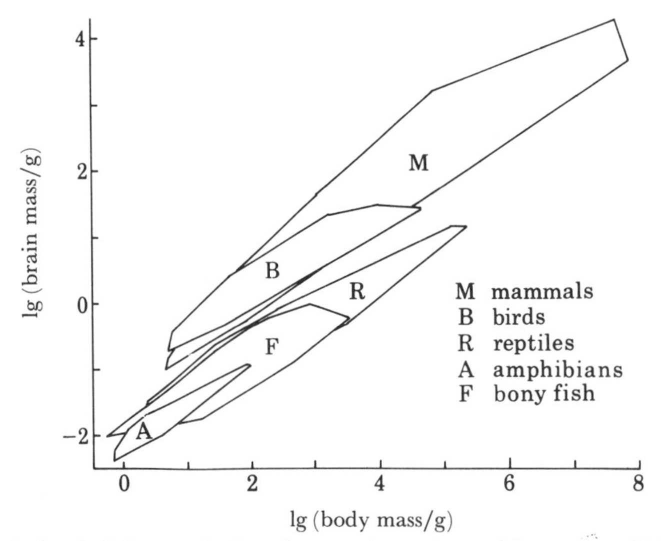

FIGURE 1. Brain–body relations in living species from five vertebrate classes. Mammals and birds from Martin (1981, 309 species of placental mammals and 180 species of birds); bony fish (Ridet 1973; 46 species), amphibians (Thireau 1975; 40 species), and reptiles (Platel 1979; 48 species). The polygons are minimum convex hulls drawn about the data of each class. See also Jerison (1973).

groups. This graph can be interpreted as a map of evolutionary opportunities in brain–body relations that have been realized in living species. It shows the regions in 'brain–body space' occupied by living species that are at various grades of encephalization and summarizes the present adaptive radiation of vertebrates. When supplemented by fossil data the graph suggests the constraints on the evolution of the brain–body system that limit the amount of brain that can evolve in a species.

The arrangement of the polygons justifies the division of the vertebrates into 'higher' and 'lower' groups with respect to encephalization. The angular orientation of the polygons, easily imagined as dispersions about regression lines, expresses the allometric factor, and the vertical displacement of the higher from the lower sets of polygons expresses the encephalization factor.

Were they added to figure 1, data on the classes Chondrichthyes and Agnatha, the cartilaginous fish and jawless fish (see, for example, Ebbesson & Northcutt 1976; Northcutt 1981), would complete this picture of the present adaptive radiation of vertebrates with respect to encephalization. The chondrichthian polygon lies between those for lower and higher vertebrates and overlaps both. Some sting rays are at a mammalian or avian grade, and that favourite 'primitive generalized vertebrate' of the comparative anatomy class, the dogfish (*Squalus*), is more encephalized than any of the 'lower vertebrates', lying near the lower bound of the higher vertebrate polygons. Sharks are, thus, intermediate rather than lower or higher. A chondrichthian polygon was omitted from the graph for didactic reasons, to maintain the clarity of the distinction of lower from higher vertebrates with respect to encephalization.

Data on living lampreys (*Petromyzon*) fall well below the 'lower' vertebrate polygon. Interpreted as an evolutionary record this could imply that the transition from an agnathan to gnathostome grade may have been the first advance in encephalization in vertebrates, since the earliest fossil vertebrates of about 450 million years ago were agnathans, and the jawed fish did not appear until about 50 million years later. Living agnathans, which are parasitic, are usually thought of as degenerate descendants of free-living species, however, and it may be that the ancestral agnathans were comparable to living fish in encephalization. The issue is of the kind that can sometimes be resolved by the fossil record.

Mappings such as those in figure 1 are the framework for interpreting an extensive fossil record on brain size, developed from the analysis of fossil endocranial casts (endocasts) for which the cranial cavity is the mould. From studies on living species we know that in birds and almost all mammals endocasts provide excellent pictures of the external surfaces of freshly dissected brains. These are less adequate in lower vertebrates, but are often good enough to enable one to estimate total brain size. Hundreds of fossil vertebrate endocasts are available for study, and together they provide a detailed record of the history and evolution of the brain (Blumenberg 1983; Edinger 1975; Hopson 1979; Jerison 1973; Radinsky 1979). Although there is occasionally some disagreement on method, it is relatively easy to estimate body size in fossil species in which enough skeletal material has been preserved. It is, therefore, possible to analyse the evolution of encephalization in fossil species by determining the extent to which their data fall into appropriate polygons of the type shown in figure 1.

In general, the differentiation between lower and higher living vertebrates as shown in figure 1 occurs in the same way in fossil species as in living species from the same groups. The most important additional information from the fossils is that early species of birds and mammals were at, or perhaps even slightly below the lower margins of the polygons for the living populations, indicating that encephalization occurred but not to the same extent as in 'average' living species. One fossil shark studied by R. Zangerl (personal communication), which lived about 250 million years ago, was probably as encephalized as living dogfish, suggesting that the sharks were the first vertebrate species to 'experiment' with encephalization as an adaptation. Other 'experiments' with enlarged brains occurred in certain dinosaurs (Ornithomimidae; see Russell 1972), and some of the mammal-like reptiles (Therapsida) may also have been encephalized beyond the present reptilian grade (Hopson 1979; Kemp 1979; cf. Quiroga 1980). The orientation of the polygons for lower and higher vertebrates when fossil data are added is the same as in living species, and the vertical displacement is similar (Hopson 1977, 1979; Jerison 1973). Such analyses lead to some straightforward conclusions, which can destroy old myths. For example, dinosaurs as a group were normal reptiles in relative brain size, and a few were even as large-brained as some living birds. The dinosaurs did not become extinct because of their small brains.

The question of the differentiation between an agnathan and gnathostome grade raised earlier is, unfortunately, not resolvable by the fossil record. I reviewed that fossil evidence some years ago (Jerison 1973) and concluded that agnathan endocasts, though impressed as beautiful patterns on the armoured fossil skulls (Stensiö 1963), were inadequate for judgements about brain size.

The orderliness of the evolution of encephalization as indicated by the analysis of fossil endocasts is impressive. I have discussed the issue in several reviews as well as in my monograph of a decade ago (Jerison 1973, 1982, 1985). With some exceptions, most other reviewers (see Armstrong & Falk 1982; Blumenberg 1983; Hopson 1977, 1979; Passingham 1982; Radinsky 1979; Tobias 1971) have also been impressed with this orderliness. The major conclusions are the following.

(i) A basal lower vertebrate grade of encephalization evolved in the earliest bony fish, amphibians and reptiles and has continued to the present as a steady-state or equilibrium maintained for a least 350 million years. Since about two-thirds of living vertebrate species are members of these three classes of vertebrates, this basal grade is the norm for vertebrates.

(ii) There are variations in encephalization within the lower vertebrate groups, the most interesting being between herbivorous and carnivorous dinosaurs. The carnivores were apparently significantly more encephalized.

(iii) The earliest fossil birds and mammals with known endocasts had evolved to a higher grade, representing at least three or four times as much brain as in lower vertebrate species of comparable body size. This progressive or 'anagenetic' evolution occurred at least 150 million years ago, and in the case of the mammals may have begun with their reptilian ancestors at least 50 million years earlier.

(iv) Within the mammals there is a good fossil record of the brain, which is consistent with a picture of steady-states punctuated by rapid evolution to higher grades. However, many grades of encephalization are represented in living mammalian species, with some (opossum, hedgehog) at the same grade as the earliest of the mammals.

(v) Two unusual conclusions are evident in the history of encephalization in primates. First, primates have always been a brainy order, perhaps doing with their brains what many other species did by morphological specializations. Second, the evolution of encephalization in the primates followed rather than preceded or even accompanied other adaptations by primates to their niches. Washburn (1978) has pointed this out as a feature of hominid evolution, but it appears to have been true for prosimians and simians as well (Jerison 1979).

(vi) The highest grade of encephalization is shared by humans and bottlenosed dolphins (*Tursiops truncatus*). The sapient grade was attained about 200000 years ago, but cetaceans may have reached their highest grade 18 million years ago.

(vii) Encephalization in the hominids is a phenomenon of the past three to five million years, and its rapidity appears to have been unique in vertebrate evolution.

(viii) These results suggest two complementary conclusions. First, the long steady-states that occurred in most groups indicate that, on the whole, encephalization was not a major element in vertebrate evolution. A particular grade of encephalization tended to be maintained once it was achieved. On the other hand, its appearance in many different and distantly related groups is evidence of some Darwinian 'fitness' for encephalization.

2. The meaning of brain size

As a first approximation, gross brain size is a kind of statistical estimator of total neural information processing capacity. The evidence is from the mammalian brain, in which the necessary quantitative analyses have been performed, but elementary ideas on how nerve cells must be packed into brains suggest that the perspective is appropriate for all vertebrates.

Brain size estimates the processing capacity of a brain because of the statistical orderliness of the brain's structure. Several aspects of that orderliness in the mammalian brain are especially relevant for information-processing. There is, first, a 'basic uniformity in the structure of the neocortex' described by Rockel *et al.* (1980), who showed that the various regions of the cortex in various mammals follow a common structural plan, with regional variations. Secondly, the neocortex is organized into columnar modules, which extend through the full depth of the cortex and are about 250 μm in diameter. Modules seem to be units of information processing (Eccles 1979; Mountcastle 1978; Szentagothai 1978), analogous to chips in computers. The number of modules is proportional to the cortical surface area, and processing-capacity should, therefore, be proportional to the total cortical surface area. For this reason a third feature of the brain's orderly structure is especially important, namely that at the between-species level the cortical surface area is determined almost entirely by brain size. The correlation (log data) is $r = 0.995$ (Jerison 1983). The syllogism is now complete: information processing capacity is determined by cortical surface area (in mammals), and cortical surface area is determined by brain size. Brain size, therefore, determines processing capacity. Since the analysis is statistical, it is more appropriate to conclude that brain size 'estimates' rather than 'determines' processing capacity.

Derived from evidence on mammals, these ideas may be extended to other vertebrates by considering the packing of neurons and glial cells into brains. Despite the obvious oversimplification, it is appropriate to assume a general statistical similarity among brain cells, at least to the extent of defining an average cell. With respect to packing, a large population of neurons and glial cells may be treated as approximately the same as an equal number of average cells. For the analysis of packing one can work with information on the total volume of an average neuron–glial unit and represent it by its equivalent sphere. The packing of such spherical model cells into a brain would then be analogous to the packing of spheres in any container. (Such a model is oddly appropriate. The mathematical solution of this packing problem is that up to about three-quarters of the space in a container can be taken up by the spheres (Sloan 1984). Evidence from the analysis of the volume of the extracellular space in the brain (Nicholson

1979) is consistent with that percentage.) Within the mammals, neuron density and the ratio of neurons to glia are both related in an orderly way to brain size, as would be expected from this geometrical model (Jerison 1985). There is no reason to assume major differences for those relations in other vertebrate classes, and the analysis is, essentially, one of parsimony. How are cells packed into their available spaces? Lacking contrary evidence, we should assume the packing is efficient, and geometrically similar for all cells in all brains, at least to the extent to permit the assumption that the number of units of information-processing in different brains is always related to the size of the brain. We may, therefore, treat the results on mammals as appropriate models for other vertebrate brains.

In the analysis of encephalization the total size of the brain is divided, statistically, into two fractions, one related to body size and the remainder (residual) to 'encephalization'. The correlation of brain size with processing capacity refers to the entire brain, however, and not solely to a component related to encephalization. The allometric and encephalization components are statistical estimators of how, not where, processing capacity is divided. It would be impossible to state, on the basis of the analysis presented thus far, whether any particular bit of brain tissue belongs to the allometric component or the encephalization component. It would be inappropriate, certainly, to think of the allometric component as sensorimotor and the encephalization component as referring to association systems if these are considered as localized grossly in different parts of the brain. In accordance with recent analyses (Diamond 1979), it is more appropriate to think of various blocks or columns of tissue as contributing to both projection and association functions, and the biometric division into allometric and encephalization factors might correspond, approximately, to a partitioning of the work of a columnar modular unit into sensorimotor and association activities.

To explain encephalization it is sometimes assumed that it is the result of reorganization of the mammalian brain that followed the evolution and subsequent enlargement of the uniquely mammalian cerebral cortex of the forebrain. This is at best an oversimplification. Many brain structures, including the forebrain, became larger in mammalian encephalization. The enlargements are correlated, and an outstanding example is the cerebellum in mammals, the size of which is closely related to the size of the forebrain. The evidence is in figure 2.

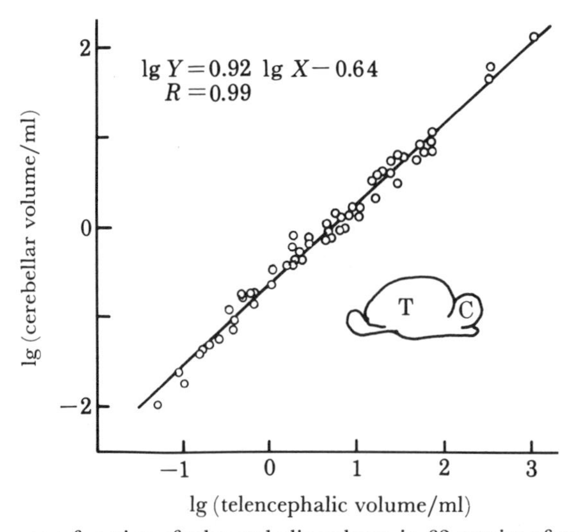

Figure 2. Cerebellar volume as a function of telencephalic volume in 63 species of mammals (24 insectivores and 59 primates). Data from Stephan *et al.* (1970). Telencephalic volume is exclusive of the olfactory bulbs. Inset sketch of mammalian brain indicates telencephalon (T) and cerebellum (C).

Figure 2 may be an example of what Mountcastle (1978) has described as the 'distributed' network in brain function, that many widely separated units contribute to the normal functions of the brain. The cerebellum is involved in conditioning and learning (McCormick & Thompson 1984), for example, indicating that it may have important functions in 'higher mental processes' beyond its traditional role as a motor control system. In any case, the enlargement of the cerebellum contributed significantly to structural encephalization in mammals. I summarized the results of multivariate analyses of structures contributing to brain function and of the correlated evolution of these structures in encephalization (Jerison 1973, pp. 63–81), and later reviews are included in Armstrong & Falk (1982).

Up to this point the brain has been treated as if it were similar in all species, and as if size alone was important in its function. This is, of course, another simplification, introduced partly because the concern here is mainly with effects of size and also because the importance of the size of biological systems, including the brain, is often under-emphasized. Brains of different species are organized differently, just as the behaviours of different species are different; the structural reorganization of the brain is consistent with the reorganization of behaviour. For example, the superior colliculi of the visual systems of reptiles and birds and some fish are so enlarged relative to the rest of the brain that they are called optic *lobes*, comparable in size to lobes of the forebrain, and sometimes larger than the entire forebrain in some species of fish and reptiles. It is more or less true that the mass of neural systems related to specialized behaviours is proportional to the importance of the behaviour for a species (cf. the 'principle of proper mass', in Jerison 1973).

Thinking statistically, we recognize gross brain size as a natural biological statistic, which estimates many significant parameters. One of these is the surface area of the neocortex in mammals, and another, at a different level of analysis, is total information processing capacity of a brain. As might be expected, brain size is a good estimator of the sizes of most of the parts of the brain, including cerebellum. Furthermore, differences in organization that occur often average out, in a sense, so that different species may be equally encephalized but for different reasons. Relatively enlarged visual system in a 'visual' species may be comparable in volume to relatively enlarged auditory system in another, 'auditory', species, and the two species could be equally encephalized despite their different specializations.

Reorganization in the brain is the basic correlate of most of the interesting variations among species in their behavioural adaptations, and the brain tissue related to these specializations could contribute to both the allometric and encephalization factors. When the specialization requires very large amounts of neural tissue it may be reflected directly in the encephalization factor. It is likely that encephalization in monkeys and apes, for example, which is about twice as great as in 'average' living mammals, results mainly from the enormous expansion of the visual system of the brain in higher primates. Equal amounts of encephalization could and normally do result from very different patterns of specialization. The extra processing capacity involved in encephalization may be allocated to very different behavioural adaptations.

We should consider now a more general question: why did encephalization evolve? It is, after all, an expensive adaptation for the economy of an organism, requiring metabolically demanding neural tissue rather than energetically 'cheaper' bone, muscle or other tissues that are required for many behavioural adaptations. The answer must be in the requirement for specialized neural control systems, and the question becomes: what kinds of neural control systems require very large amounts of tissue?

The evidence on mammalian brains is that their enormous masses are involved mainly in the works of the projection systems (Diamond 1979; Merzenich & Kaas 1980). The information that they process would be characterized as sensory and motor. This suggests that the work of the brain involves perceptual systems, which operate on sensory and motor information that is transmitted in the nervous system. The amount of information is enormous, involving thousands of millions of changes of state every second, which must somehow be related to the changing information received through sensory and motor surfaces about the external environment, often by a moving animal. In reviewing the problem (Jerison 1973) I thought that Craik's (1943) discussion was helpful, that one could think of the work of the brain as constructing a model of a possible world in which the changing pattern of the detailed neural information that is being processed finds its place in an understandable way. In modern terms this would be described as a hierarchically organized system, in which events such as the activity of single neurons, or even modular columns of neurons, are organized at lower levels of the hierarchy, and the 'possible world' is represented at higher levels as part of a more complex model. The complex model, in fact, is a real world of ordinary experience containing, for example, 'objects' in 'space' and 'time'. Objects, space and time are chunks of information, as it were, more useful than the elementary data (Simon 1974).

Gross brain size may, thus, be thought of as having two kinds of meaning. First, it is the morphological measure of neural information processing capacity, reflecting the orderly packing of processing units within brains. Second, it is a quantity which, upon reaching some particular size and handling some specifiable amount of information, handles that information by creating a representation of a possible world, within which the patterns of neural activity make sense. This second meaning is, in fact, nothing less than the creation of a reality, of a real world within which the events of a lifetime take place. Familiar computer analogies would have 'pattern recognition' as part of that work, although it would be more accurate to call it 'pattern construction'. The patterns created by a brain would be much more interesting, with many more dimensions than the visual and auditory patterns created by computers. The brain's patterns have emotional and motivational dimensions, anticipatory dimensions, and even dreams and hallucinations. For our purpose, however, the point to emphasize is that it is functions such as these that require the enormous amounts of tissue that are in the brain.

3. Intelligence

Encephalization was discovered in the search for neurological correlates of intelligence, and it is easier to analyse encephalization. It has an identifiable evolutionary history and unusual but interesting features as a 'character' under natural selection. Animal intelligence and the problems in its analysis may be much better understood if we reverse the implicit causal arrow and think of it as the behavioural correlate of encephalization. The evidence of encephalization is so clear that were we ignorant of behaviours that might be correlated with it, or were we lacking in intuitions about the behaviours, we would find it necessary to invent a category of behaviour related to encephalization.

Evolutionary traits are often analysed for their phyletic histories to determine 'cladistics' and lineages. Although this can be done for encephalization in the hominid lineage, it is not an easy exercise for other animal groups (cf. Radinsky 1979). There is no feature in its definition that would require a taxonomic label to help describe a grade of encephalization that is identified in a species. Deer and wolves are equally encephalized, for example; they are average mammals joined by lemurs among the primates and even by crows. Encephalization is a measure of capacity, independent of the way the capacity is used, and it may be treated as a trait for 'anagenetic' rather than 'cladistic' analysis (Gould 1976). Deer and wolves and lemurs and crows are more or less equal on this quantitative trait.

This provides the first insight about animal intelligence. When distantly related animal species are comparable in excess neural information processing capacity, that is, in encephalization, we should identify the species as comparable in intelligence. Yet the near equality in encephalization may be based on radically different adaptations. Some species of deer and wolves (to retain that example, useful in many contexts in this discussion) live in ecological balance, as prey and predator species (Mech 1970), and the fact of the balance, which depends on appropriate behavioural adaptations, suggests that they should be equated behaviourally despite the different repertoires of the species. In that sense their approximate equality in encephalization reflects their behavioural grades and is useful as part of the characterization of each species.

The insight is that comparable *amounts* of intelligence in different species may not (and normally would not) reflect comparable *kinds* of intelligence. Many and various intelligences (in the plural) must have evolved in conjunction with evolving environments and with brains and behaviours adapted to those environments.

That intelligences would be of various kinds is almost an axiom of evolutionary analysis, since adaptations evolve in the contexts of the environments in which they are effective, and species never occupy identical niches. The evolution of neural and sensorimotor adaptations provides many fine examples of uniqueness of species. The visual systems of deer and wolf, for example, may be similar in many ways, for example, in the structure of the sensory cells, neural networks of the retina, and the central nervous pathways and centres (cf. Merzenich & Kaas 1980). Yet these systems are significantly different: the deer, like most ungulate 'prey' species, probably has panoramic vision whereas the wolf's visual field is more nearly like the primate's proscenium stage (cf. Marler & Hamilton 1966). The visual system encumbers significant amounts of nervous tissues and, thus, contributes to brain size and measured encephalization. Neural machinery associated with the sensory systems and motor control systems as a group determines a large fraction of the mass of the whole brain. Equality of encephalization of deer and wolf, thus, implies that the neural control systems for the specialized adaptations, though different in the two species, sum to approximately equal amounts relative to body size.

Laboratory scientists should anticipate difficulties when attempting to compare species in intelligence, because it may be impossible, even in principle, to equate the environments used in testing different species. Behaviours and their control systems evolved in specific environmental contexts, as adaptations to specialized environments . The difference between 'intelligent' and 'unintelligent' behavioural adaptations is in the amount of neural tissue that they encumber, according to the present perspective, which may be uncorrelated with measurable differences in overt behaviour. Intelligence as a correlate of encephalization would be manifested in experience, rather than behaviour, in the realities created in the brain of a species, and although this view has its charms it obviously adds significantly to the difficulties of a scientific analysis requiring objective tests.

The first inference was, thus, that a variety of intelligences may be represented at a single grade of encephalization. A second inference from treating intelligence as the behavioural counterpart of encephalization is about the class of behaviours that is likely to be involved in the evolution of intelligence. Ever since Darwin and Romanes it has been normal to assume that 'brainy' behaviours involve the ability to profit from experience, the ability to learn. An enormous literature now indicates that animals of many different species can usually learn a particular task, and apparent differences among species in learning ability are determined as much by the skill of the animal trainer as by anything else. Operant conditioning procedures have been remarkably successful in training animals to act as if their performance was based on 'higher mental processes', even when it was demonstrably based on associative learning. The evidence is on the learning abilities of pigeons and other unencephalized species. Macphail (1982), Warren (1977) and many others have reviewed the sorry record of research correlating learning ability with 'phylogenetic level of advancement'. There is at best equivocal evidence of an orderly progression in the capacities of different species for conditioning and associative learning that corresponds with their degrees of encephalization (Bitterman 1975; Riddell 1979). More significantly, it is not clear that any of the simpler categories of learning are dependent on large amounts of nervous system, and the evolution of learning ability is not implicit in the evolution of encephalization.

There is empirical evidence for a relation between the grade of encephalization and what an animal appears to know when it copes with a task. This at least seems to be the case for learning sets performance in primates (Passingham 1982) and for Piagetian 'conservation of mass' as demonstrated in a chimpanzee (Premack & Woodruff 1978). Such results are entirely consistent with the view presented earlier about the kinds of brain functions that require very large amounts of brain tissue for their control. As pointed out in the previous section, the enlargement of vertebrate brains, especially the very great enlargement of the mammalian brain, can be related to the evolution of extensive sensorimotor 'projection' systems for all of the sensory modalities and for motor feedback. And the enlargement is involved not merely in the use of sensorimotor information but in the construction of representations of reality from neural data. The 'intelligence' that corresponds to higher grades of encephalization is one involving a knowledge of reality, or, in terms of the earlier discussion, the quality of the reality created by the brain to account for the information that is received.

I should, perhaps, make the point that the view of a brain as creating reality is not solipsistic. It does not deny an external reality. The fact of an external reality explains the uniformities in social behaviour and shared experiences. There are fundamental similarities among the operations of different nervous systems, and there are constancies in the environments to which nervous systems are exposed. The realities created by brains reflect these constraints. They are not chaotic but must be similar for different species and very similar for individuals of the same species. Your reality and mine are similar enough to leave little doubt that we share a view of a real external world that remains constant as we live our lives. Our perceptions and experiences are referred to that reality, and it is only in our more metaphysical moments that we question it.

Grades of encephalization presumably correspond to grades of complexity of information processing. These, in turn, correspond in some way to the complexity of the reality created by the brain, which may be another way to describe intelligence. There are problems here in taking evidence of the presence of unusual kinds of information processing as implying that the processing is difficult and can only be accomplished with very large amounts of brain tissue and the conclusion that this must be an example of complex processing. The issue arises in analysing exotic adaptations, of which echo-location is a good example.

Echo-location and sound-ranging in insect-eating bats and in toothed whales are adaptations for handling spatial information, which depend on the activity of an auditory–vocal system. The adaptation clearly requires a fair investment of neural machinery. Although much of the bat's brain is devoted to this class of information, the whale's investment must exceed the bat's by an enormous amount. It is sometimes suggested that the size of the brain in large whales, which is several times the size of the human brain, is a result of the processing capacity required for this exotic adaptation for handling information. But if it were merely a problem of handling information, whales could do the job with a much smaller investment in nervous tissue; they would need no more than the amount of tissue in a bat's brain. There must be more than echo-location that explains encephalization in whales, in which dolphins and killer whales are at approximately the human grade. We must infer that the reality created from this information in the whale is more elaborate, more complex, and at least very different from that produced by the bat's brain, that the realities of these species differ radically.

To extend the analogies we can compare human language as a sensory–vocal adaptation with sensory–vocal adaptations in other primates, indeed in other mammals. Specialized processing of auditory information and of vocalization is reasonably similar in different species, including the human species. The important difference is obviously in the way the information contributes to the realities that we construct. Our linguistic world is a unique reality-creating world, which is dramatically demonstrated by the facts of literacy: reading can enable one to enter a fictional world and live in it. We create worlds with language. The acoustic story of language is limited and is not much different from the acoustic story of vocalization and hearing in other species, but the cognitive story is another matter.

This leads to a final inference from encephalization as defining animal intelligence, an extension of the proposition that enlarged brains are enlarged because of their activity in the sensory–motor and perceptual domains of analysis. The inference might be stated more clearly by calling this domain by a better name, the cognitive domain. The idea is inherent in the hierarchical organization of large information handling systems, including large brains. In this view, following Craik's (1943) perspective as well as Simon's (1974), there is a major problem of organizing the information. The solution is by chunking, nesting subroutines within larger subroutines. The place of representation of the external world, the creation of a reality, in this scheme is as a method to make sense of an otherwise impossible amount of information. In ordinary experience some of this creation is consciousness or awareness of the external world, which we recognize as a simplification when we examine any part of that world carefully. There is always more detail revealed in careful examination and our knowledge expands as the examination continues. The elements of the world are the 'chunks' and they may have the form of objects, or of dimensions, or of any category within which experience is organized. The model of reality is made simple or elaborate, depending on the observer's requirements and those imposed by the information that is being handled.

This leads to the expectation that very encephalized animals may be something like us in having unusual adaptations for handling or constructing realities. Perhaps language is the most complex of human adaptations, and some of the unusual features of language as a medium of communication are clearly related to its role in the creation of a reality. In the following discussion let us keep in mind that the idea of creating reality is a dramatic way of describing the 'knowing' of reality, or cognition, and that this is a statement about neural activity at a high level of hierarchical organization. Although theoretical, based on word games, this is a *biological* rather than philosophical, psychological, or linguistic approach to language. Let us also note that the analysis of brain size indicates that it is only systems that ultimately contribute to cognitive functions, the knowing of reality, that account for very large amounts of nervous system, and that the very extensive representation of language in the brain argues for its role as a cognitive system.

The understanding of language in this biological perspective may be different from the usual understanding. Language is usually equated with communication, and discussions of animal communication and animal language are treated as equivalent. This may be a serious mistake that can lead to great confusion in the understanding of both language and communication. Here is my analysis.

Human language as a cognitive adaptation contributes to the reality constructed by each individual. In this respect it is like vision or olfaction or other senses. But language is also a major adaptation for communication. The implication is that we communicate by sharing parts of our real worlds, sharing consciousness with one another, because we share part of the information that we use in constructing our realities. We can verify this by the often repeated 'experiment' in which we have all participated and which I described earlier. When we read and become engrossed in a realistic novel, we experience the realities of the characters as vividly as if we were living their lives. This verifies the fact that when we communicate with language we share experiences, and this is an unusual way for animals to communicate.

It is as if we could communicate by having others see what we see and hear what we hear. The role of language in communication is very close to fictional accounts of communication by extrasensory means and may explain the attractiveness of ideas of such psychic powers. These imagined powers are not far removed from what we do in everyday life when we use ordinary language. But the penalty for our exotic method of communication is uncertainty about the information, misunderstanding and false understanding.

Normal animal communication, we have learned from the ethologists, is direct and certain. It is with commands that are usually obeyed: *sign stimuli* and *releasers* and *fixed action patterns*. We can use elements of human language to train other animals, and there is some suggestion that when this has been done with great apes and with dolphins, cognitive activity becomes involved in the communication (Premack & Woodruff 1978; Herman 1980; Herman *et al.* 1984). But there is as yet no evidence that the communication is in any sense like human communication with shared consciousness. Rather the homologue for animal communication in the human species may be the important communication with 'natural' gestures, such as unrehearsed facial expressions, which might be described as communicating directly without the intervention of consciousness or a sense of identification. It is also sometimes described as limbic language (Myers 1976), and it has some of the characteristics of animal communication in its universality and lack of ambiguity.

I have tried in these remarks to suggest the nature of the evolution of intelligence as an aspect of the evolution of encephalization. This has led to a definition of intelligence as processing capacity beyond that required for routine bodily functions. It led also to the assertion that the 'excess' capacity (in Lashley's words) is used primarily for the construction of reality: the representation of a world that is the reality of each species. In my final example I indicated that human reality is deeply associated with human language. It is reasonable to extend this implication to the specialized correlates of encephalization in other species, and to suggest that their adaptations may be as unusual as language. The correct view is surely that in the evolution of excess processing capacity, that is, in the evolution of encephalization, a variety of intelligences evolved. Human intelligence, deeply associated with human language, is one kind of intelligence. The evolutionary message about intelligence, like the message about so many other dimensions in biology, is a message about pluralism and diversity, about the variety of intelligences in the biological world.

References

Armstrong, E. & Falk, D. (ed.) 1982 *Primate brain evolution: methods and concepts*. New York and London: Plenum.

Bitterman, M. E. 1975 The comparative analysis of learning. *Science, Wash.* **188**, 699–709.

Blumenberg, B. 1983 The evolution of the advanced hominid brain. *Current Anthrop.* **24**, 589–623.

Craik, K. J. W. 1943 *The nature of explanation*. London and New York: Cambridge University Press. (Reprinted in 1967, with postscript.)

Diamond, I. T. 1979 The subdivisions of the neocortex: A proposal to revise the traditional view of sensory, motor, and association areas. *Prog. Psychobiol. & physiol. Psychol.* **8**, 1–43.

Ebbesson, S. O. & Northcutt, R. G. 1976 Neurology of anamniotic vertebrates. In *Evolution of brain and behavior in vertebrates* (ed. R. B. Masterton, M. E. Bitterman, C. B. G. Campbell and N. Hotton), pp. 115–146. Hillsdale, New Jersey: Erlbaum.

Eccles, J. C. 1979 *The human mystery*. New York: Springer-Verlag.

Edinger, T. 1975 Paleoneurology, 1804–1966: An annotated bibliography. *Adv. Anat. Embryol. Cell Biol.* **49**, 12–258.

Gould, S. J. 1976 Grades and clades revisited. In *Evolution, brain and behavior: persistent problems*. (ed. R. B. Masterton, W. Hodos and H. J. Jerison), pp. 115–122. Hillsdale, New Jersey: Erlbaum.

Herman, L. M. 1980 Cognitive characteristics of dolphins. In *Cetacean behavior: mechanisms and functions* (ed. L. M. Herman), pp. 363–429. New York: Wiley.

Herman, L. M., Richards, D. G., & Wolz, J. P. 1984 Comprehension of sentences by bottlenosed dolphins. *Cognition* **16**, 129–219.

Hopson, J. A. 1977 Relative brain size and behavior in archosaurian reptiles. *A. Rev. Ecol. Systematics* **8**, 429–448.

Hopson, J. A. 1979 Paleoneurology. In *Biology of the Reptilia* (ed. C. Gans, R. G. Northcutt and P. Ulinski), vol. 9, pp. 39–146. London and New York: Academic Press.

Jerison, H. J. 1973 *Evolution of the brain and intelligence*. New York: Academic Press.

Jerison, H. J. 1979 Brain, body, and encephalization in early primates. *J. Human Evol.* **8**, 615–635.

Jerison, H. J. 1982 The evolution of biological intelligence. In *Handbook of human intelligence* (ed. R. J. Sternberg), pp. 723–791. New York and London: Cambridge University Press.

Jerison, H. J. 1983 The evolution of the mammalian brain as an information processing system. In *Advances in the study of mammalian behavior* (ed. J. F. Eisenberg & D. G. Kleiman), pp. 113–146. Special publication no. 7, American Society of Mammalogists.

Jerison, H. J. 1985 Issues in brain evolution. *Oxford surveys evolut. biol.* **2**. (In the press.)

Kemp, T. S. 1979 The primitive cynodont *Procynosuchus*: functional anatomy of the skull and relationships. *Phil. Trans. R. Soc., Lond.* B **285**, 73–122.

Lashley, K. S. 1949 Persistent problems in the evolution of mind. *Q. Rev. Biol.* **24**, 28–42.

Macphail, E. M. 1982 *Brain and intelligence in vertebrates*. Oxford: Clarendon.

McCormick, D. A. & Thompson, R. F. 1984 Cerebellum: essential involvement in the classically conditioned eyelid response. *Science, Wash.* **223**, 296–299.

Marler, P. & Hamilton, W. J. III. 1966 *Mechanisms of animal behaviour*. New York: Wiley.

Martin, R. D. 1981 Relative brain size and basal metabolic rate in terrestrial vertebrates. *Nature, Lond.* **293**, 57–60.

Martin, R. D. 1983 *Human brain evolution in ecological context*: James Arthur Lecture on the Evolution of the Human Brain. American Museum of Natural History.

Mech, L. D. 1970 *The wolf: the ecology and behavior of an endangered species*. New York: Natural History Press.

Merzenich, M. M. & Kaas, J. H. 1980 Principles of organization of sensory-perceptual systems in mammals. *Prog. Psychobiol. physiol. Psychol.* **9**, 1–42.

Mountcastle, V. B. 1978 An organizing principle for cerebral function: The unit module and the distributed system. In *The mindful brain* (ed. G. M. Edelman and V. B. Mountcastle), pp. 7–50. Cambridge, Massachusetts: M.I.T. Press.

Myers, R. E. 1976 Comparative neurology of vocalization and speech: proof of a dichotomy. *N.Y. Acad. Sci.* **180**, 745–757.

Nicholson, C. 1979 Brain-cell microenvironment as a communication channel. In *The neurosciences: fourth study program* (ed. F. O. Schmidt & F. G. Worden), pp. 457–476. Cambridge, Massachusetts: M.I.T. Press.

Northcutt, R. G. 1981 Evolution of the telencephalon in nonmammals. *A. Rev. Neurosci.* **4**, 301–350.

Passingham, R. E. 1982 *The human primate*. San Francisco: Freeman.

Platel, R. 1979 Brain weight–body weight relationships. In *Biology of the Reptilia* (ed. C. Gans, R. G. Northcutt and P. Ulinski), vol. 9, pp. 147–171. London and New York: Academic Press.

Premack, D. & Woodruff, G. 1978 Does the chimpanzee have a theory of mind? *Behav. Brain Sci.* **4**, 515–526.

Quiroga, J. C. 1980 The brain of the mammal-like reptile *Proainognathus jenseni* (Therapsida, Cynodontia). A correlative paleo-neurological approach to the neocortex at the reptile-mammals transition. *J. Hirnforschung* **21**, 299–336.

Radinsky, L. 1979 *The fossil record of primate brain evolution.* The James Arthur Lecture. New York, American Museum of Natural History. 27 pp.

Riddel, W. I. 1979 Cerebral indices and behavioral differences. In *Development and evolution of brain size: behavioral implications* (ed. M. E. Hahn, C. Jensen and B. C. Dudek), pp. 89–109. New York: Academic Press.

Ridet, J.-M. 1973 Les relations pondérales encephalo-somatiques chez les poissons téléostéens. *C. r. hebd. Séanc. Acad. Sci., Paris* 276: 1437–1439.

Rockel, A. J., Hiorns, R. W. & Powell, T. P. S. 1980 The basic uniformity in structure of the neocortex. *Brain* **103**, 221–244.

Russell, D. A. 1972 Ostrich dinosaurs from the Late Cretaceous of Western Canada. *Can. J. Earth Sci.* **9**, 375–402.

Simon, H. A. 1974 How big is a chunk? *Science, Wash.* **183**, 482–488.

Sloan, N. J. A. 1984 The packing of spheres. *Scient. Am.* **250**, Jan., 116–125.

Stensiö, E. 1963 The brain and cranial nerves in fossil, lower craniate vertebrates. *Skr. Nor. Videnskaps-Akad. Oslo, Mat. Naturv. Kl.* (N.S.) no. 13.

Stephen, H., Bauchot, R. & Andy, O. J. 1970 Data on size of the brain and of various parts in insectivores and primates. In *The primate brain* (ed. C. R. Noback & W. Montagna), pp. 289–297. New York: Appleton-Century-Crofts.

Szentagothai, J. 1978 The neuron network of the cerebral cortex: A functional interpretation. *Proc. R. Soc. Lond.* B **201**, 219–248.

Thireau, M. 1975 L'allométrie pondérale encephalo-somatique chez lez urodeles. II. Relations interspécifiques. *Bull. Mus. nat. Hist. natur.* **207**, 483–501.

Tobias, P. V. 1971 *The brain in hominid evolution.* New York: Columbia University Press.

Warren, J. M. 1977 A phylogenetic approach to learning and intelligence. In *Genetics, environment and intelligence* (ed. A. Olivero), pp. 37–56. New York: North Holland/Elsevier.

Washburn, S. L. 1978 The evolution of man. *Scient. Am.* **239**, Sept., 146–154.

Discussion

H. B. Barlow, F.R.S. (*University of Cambridge, U.K.*). If, as you suggest, the encephalization index is related to information processing capacity or intelligence, then the following curious paradox arises. In a heavy brain weighing hundreds of grams a high index might correspond to the addition of many grams of brain tissue, and a correspondingly large number of brain cells, whereas in a light brain of, say, 10 g the same increase of the index would correspond to the addition of a comparatively small number of cells. Why does the heavy brain require many times as many extra cells as the light brain to confer the same increase in intelligence?

H. J. Jerison. The question is fundamental. To answer we must review the units to be used in assessing neural information processing and how these would measure encephalization. The implicit premise in the question is that the unit is the nerve cell. In that case, encephalization could be measured by the number of 'extra neurons' after accounting for the number of neurons encumbered by routine body functions. A large animal species requires a larger brain than a small animal species for handling routine body functions (much as it requires a larger heart or liver), but when the larger-bodied species is encephalized to the same extent as the smaller one, it should have the same number of 'extra neurons' as the smaller species. The fractional enlargement of the brain due to encephalization (beyond the 'allometric enlargement associated with body size) would have to be greater in the larger species because there are fewer neurons per unit volume in larger than in smaller brains. The nonparadoxical answer is, thus, that the increase in brain size related to intelligence could be based on the same number of nerve cells in large and small brained species, and that size differences despite equal encephalization in large and small brains are due to the way nerve cells are packed in brains (for quantitative examples, see Jerison 1963).

An alternative processing 'unit' might be a brain map of a peripheral sensory or motor surface. Equally encephalized species should then have comparably extensive mappings in their brains, but to maintain constant ratios of central brain cells to peripheral body cells the brain maps in larger species would have to contain more neurons. (This is one explanation for brain–body allometry.) Advances to higher grades of encephalization would be by increasing the number or complexity of the maps. Such increments would require multiplication rather than addition of cells, and *more* neurons for equal encephalization in larger relative to smaller brains. Both an additive 'extra neurons' factor and a multiplicative mapping factor have to be considered in a complete analysis of encephalization (see Jerison 1977).

References

Jerison, H. J. 1963 Interpreting the evolution of the brain. *Hum. Biol.* **35**, 263–291.
Jerison, H. J. 1977 The theory of encephalization. *Ann. N.Y. Acad. Sci.* **299**, 146–160.

Phil. Trans. R. Soc. Lond. B **308**, 37–51 (1985) [37]
Printed in Great Britain

Vertebrate intelligence: the null hypothesis

By E. M. Macphail

Department of Psychology, University of York, Heslington, York YO1 5DD, U.K.

Human cognitive capacities have so evolved that man is able to solve an extensive range of problems having very different properties. Comparative psychologists have endeavoured to throw light on the evolution and nature of this general intellectual capacity by exploring performance of non-human vertebrates in a variety of learning tasks, in the expectation of demonstrating superior intelligence in species more closely related to man. It has, however, proved difficult to establish that any observed difference in performance is due to a difference in intellectual capacity rather than to a difference in such contextual variables as, for example, perception or motivation. Three hypotheses that might account for the lack of experimentally demonstrable differences in intelligence amongst non-humans are discussed. The first proposes that the data currently available may have been misinterpreted: that, for example, the potential role of contextual variables has been exaggerated. According to the second hypothesis, the questions posed by comparative psychologists have been inappropriate: learning mechanisms are adaptations evolved for life in a specific ecological niche, so that mechanisms available to species from different niches are not properly comparable. It is argued that neither of these two hypotheses receives convincing empirical support. A third hypothesis proposes that there are, in fact, neither quantitative nor qualitative differences among the intellects of non-human vertebrates. It is argued that this null hypothesis is currently to be preferred, and that man's intellectual superiority may be due solely to our possession of a species-specific language-acquisition device.

1. Comparative investigations of intelligence

The human mind is capable of solving an essentially infinite variety of problems, problems that may be not only novel to the individual but novel also in the sense that they are problems of a type that have not been encountered until relatively recent times by any human being. The specific environmental pressures that shaped the development of the human intellect have, then, led to the evolution of a species of intelligence having very general application. Among the questions that psychologists with an interest in human intelligence seek to answer are, first, how did this general problem-solving capacity evolve, and second, what is the nature of this capacity? These two questions are not entirely independent: theories of the evolution of human intelligence must inevitably embody assumptions about its current nature and, similarly, evidence of the nature of the intellect at an earlier stage of development would carry implications for its nature at more advanced stages.

Since human beings can solve a far wider range of problems than any non-human species, humans are, in the sense in which I am using the word intelligent, more intelligent than non-human animals. It has also widely been supposed that differences in intelligence can be seen between non-human groups of animals, and that there is a ranking of species in intelligence, having humans at the highest point. Indeed some believe that there is good scientific support for this view. Jensen, for example, writes: 'in terms of measured learning and problem-solving capacities, the single-cell protozoans (for example, the amoeba) rank at the bottom of the scale, followed in order by the invertebrates, the lower vertebrates, the lower mammals, the primates, and man. The vertebrates have been studied most intensively and show fishes at the bottom of the capacity scale, followed by amphibians, reptiles, and birds. Then comes the mammals, with rodents at the bottom followed by the ungulates (cow, horse, pig, and elephant, in ascending order), then the carnivores (cats and dogs), and finally the primates, in order: new world monkeys, old world monkeys, the apes (gibbon, orangutan, gorilla, chimpanzee), and, at the pinnacle, humans' (Jensen 1980, p. 175). Now if such a ranking does exist (whether or not the ordering is that posited by Jensen), there is clearly an opportunity available to explore the evolution of intelligence: although no living species is an ancestor of man, some at least are, phylogenetically, close relatives, and it would not be unreasonable to hope to trace the evolution of the intellect by contrasting the intellectual capacities of species progressively less closely related to man. It seems inevitable that such an enquiry, if successful, would throw considerable light on the structure of the human intellect.

The widespread acceptance among non-specialists of the notion that some non-human animals (rhesus monkeys, say) are more intelligent than others (goldfish, for example) is, no doubt, based on relatively informal observations of their behaviour. For analytical studies, we turn to the work of those comparative psychologists who have attempted to deduce specific differences in intellectual function from controlled observation of performance of various species in problem-solving tasks. One broad distinction that may be made among their proposals for differences in intellect is that between qualitative and quantitative differences. By a qualitative difference between species is meant the possession by one species of a mechanism that is absent in another; *prima facie* evidence of a qualitative difference might be the observation that there was some task or set of tasks that could be solved by one species, but not by another. A quantitative difference between two species would mean that one species used a mechanism or mechanisms common to both species more efficiently than the other, and this might be reflected in a faster rate of solution or better asymptotic performance level by one species in some task solved by both.

The fact that the performance of two species in some learning task differs does not, of course, necessarily imply a difference in intelligence between the species. Before any such conclusion could be reached, plausible alternative accounts in terms of what Bitterman (for example, 1965) has called 'contextual variables' would have to be ruled out. It might, for example, be the case that the sensory or motor capacities of the species differed, or that the reward used was less effective in one species (perhaps due to a difference in motivation). It is clearly not possible to equate such variables across species, and Bitterman has recommended the use of the technique of systematic variation to rule out explanations in terms of contextual variables. This technique involves repeated tests of the species in a given task, using different levels of plausibly relevant variables: if one species, for example, mastered a food-rewarded task at any of a wide range of levels of food deprivation, and another failed at all levels, it might reasonably be assumed that the performance difference was not attributable to differences in deprivation.

What interspecies differences in intellect have, then, been proposed by comparative psychologists, and how well are these proposals supported by evidence? A survey of the literature on vertebrates (Macphail 1982) generated what were, to the author at least, somewhat unexpected answers to those questions. First, remarkably few specific proposals have been made concerning the nature of differences in intellect between vertebrate groups; second, none of the proposals applicable to vertebrates (excluding man) appears to enjoy convincing support. There have, of course, been many reports of differences between species in performance on a variety of learning tasks, and a number of these reports have indeed generated proposals for differences in intellect. My survey concluded, however, that in no case was it possible to rule out the possibility that the performance difference was due to effects of contextual variables. This may, at least in part, reflect the fact that despite the evident force of the argument for its use, the technique of systematic variation has rarely been used in practice, presumably because of its time-consuming nature.

If comparative psychology has failed to demonstrate the differences in intellect that were so confidently expected, the question that inevitably arises is, why have no differences been identified? This paper will consider three contrasting responses to that question.

2. The role of contextual variables

One response to the question might, of course, simply be that my 1982 account misinterpreted the status of current proposals (or of some of them). It could be that the evidence in their support is stronger than was alleged, that some evidence was, perhaps, unreasonably dismissed, or that other evidence was overlooked entirely. Evidently it would not be possible fully to counter such possibilities without detailed discussion of individual issues. There is, however, one general objection to the approach adopted which may appropriately be discussed here. It might be argued that an exaggerated prominence was given to the role of contextual variables and that to adopt the position that any observed performance difference should be ascribed (unless proved otherwise) to differential effects of contextual variables, makes it systematically impossible that any performance difference would be interpreted as reflecting an intellectual difference. There is no finite catalogue of potentially relevant contextual variables: how, therefore, could their effects be conclusively ruled out?

As applied to proposals for qualitative differences, this objection has little force. Most such proposals have in fact been rejected because of the successful demonstration of some phenomenon in a species previously believed incapable of it. In many cases, it has been possible also to point to contextual variables critical to the appearance or not of a phenomenon in a species. For example, an extensive series of reports using various teleost species failed to observe any improvement (such as is seen in rats) in performance across a series of reversals of simultaneous discriminations, and this led Bitterman (1965) to propose that this task reflected a qualitative difference between the rat and the fish intellect. Many subsequent studies have, however, succeeded in showing significant serial reversal improvement in fish, and it is clear that one important factor favouring its appearance is proximity of choice and reward site (see, for example, Mackintosh & Cauty 1971; Engelhardt *et al.* 1973). In fact it appears (Macphail 1982) that there is currently no phenomenon of learning demonstrable in one (non-human) vertebrate species that has not been found in all other vertebrates in which it has been sought systematically: by which is meant, of course, sought in more than one or two isolated investigations. In other words, it is simply not the case that qualitative differences in performance have been 'explained away' by reference to contextual variables.

The argument is no more convincing when applied to the rejection of proposals for quantitative differences, proposals based on quantitative differences in learning rate or asymptotic performance levels. The principal difficulty with these proposals is that there is not, for any species, any unique or standard performance level in a given type of task. The performance of a species inevitably alters as contextual variables alter, and overlap between different species has been found at certain values of such variables in most of the tasks claimed to reflect quantitative differences; in the case of other tasks, it has been possible to point to some contextual variable known to affect performance in one species, and which has not been varied systematically for the other species concerned (Macphail 1982). Moreover, in some cases performance of some supposedly inefficient species has subsequently been observed at so high a level that no further appeal to contextual variables is necessary, there being, in effect, no difference to explain. Mackintosh, for example, has suggested (for example, Mackintosh 1970) that probability learning is a task in which pigeons perform less efficiently than rats, this inefficiency reflecting a lower stability of attention in birds than mammals. But Macphail & Reilly (1983) found that some pigeons achieved perfect performance (100% choice of the majority stimulus) in a conventional probability task in which the majority stimulus was rewarded on 70% of trials (the minority stimulus being rewarded on 30% of trials). The same birds achieved this optimal performance in both colour-relevant and position-relevant versions of the discrimination (so that in at least one version, a salient irrelevant stimulus may be assumed to have been present).

A further objection might also then be made to the dismissal of the many instances of species contrasts for which no evidence derived from the use of the technique of systematic variation is available. Given that that is the only decisive evidence, and that it is unavailable, it might be argued that it is equally valid to conclude that some given performance difference is due to an intellectual contrast as to some other contrast. But this argument is clearly unsound: there is ample evidence that contextual variables do affect performance levels within a species (and in an individual animal), so that parsimony clearly requires that differences should in the first place be ascribed to them rather than to intellectual differences of which, it is argued here, there is as yet no undisputed demonstration available. There will be an opportunity to judge a specific argument in which appeal is made to contextual variables in a later section of this paper, in which the maze-learning capacities of the pigeon are discussed.

3. Ecological considerations

A second response to the absence of demonstrations of interspecies intellectual differences might be that comparative psychologists have failed because they have made invalid theoretical assumptions in designing their experiments, so that those experiments have not been appropriate to the questions of central interest. Comparative psychologists have conducted laboratory experiments in the expectation of obtaining differences in performance from distantly related groups of animals. But each living species has evolved to cope with the specific demands of the type of environment, the ecological niche, that it occupies. Animals should be selected for comparison, not because they are distantly related, but because they occupy contrasting ecological niches; animals may be distantly related but have evolved, in parallel and independently, capacities that are adapted to some similar environmental demand. According to this view, general intelligence is a misleading concept: different problems in natural environments are solved by different, problem-specific, devices, and problems in artificial laboratory environments will be solved, if they are solved at all, by devices pressed into service in tasks that at best could only give a misleading view of the true capacities of an animal.

If most problems are solved by devices that are both species- (or niche-) and problem-specific, then it would clearly not be appropriate to compare intelligence across species, nor would it be sensible to attempt to trace the evolution of a general problem solving capacity (since none exists). It is therefore important to emphasize that the claim that problem solving is achieved by a number of specific devices is an empirical claim and not, for example, a necessary consequence of acceptance of the principle of evolution by natural selection. Many general process learning theorists (for example, Dickinson 1980) now incline to favour the view that association formation may properly be seen as a process of attributing causes to events. Since there are causal links between events in all environments, it is not at all unreasonable to expect that at least some learning mechanisms might be held in common by species from very different niches. Whether mechanisms of association formation might be all that are required for all non-human forms of problem solving is a large and important question, but not directly relevant to our interest in interspecies comparisons.

As a preliminary to discussion of the 'ecological' view, it should be pointed out that one difficulty is, of course, that it still does not explain why psychologists have failed to detect differences in capacity using their artificial tasks. Goldfish, pigeons, rats and monkeys appear, at least on the surface, to occupy very different niches: surely at least one of the tasks used should have engaged a very different device in one species from that used by some other: but the problem is that no such differences have been demonstrated.

Two specific predictions follow from the position under consideration. The first is that within one species, it should be possible to demonstrate that learning in one type of task proceeds in some way differently from that seen in some other types of task. The second prediction is that, by considering tasks that exploit specific differences in the environments of two species, it should be possible to demonstrate between-species differences in learning which can plausibly be attributed to differences in learning mechanisms.

(a) *Biological constraints*

The proposal that there should be within-species differences in learning according to the type of task encountered has been intensively investigated over recent years by psychologists interested in what have become known as 'biological constraints' on learning. Such constraints would, it was believed, undermine traditional general process theories, which had assumed an interchangeability of different categories of stimuli and reinforcers in the formulation of laws of learning. As examples of such proposed constraints, it may be constructive to consider two instances drawn from poison aversion learning. Garcia & Koelling (1966) found that rats showed better conditioning of internal (taste) stimuli as compared to external (auditory and visual) stimuli when illness was the reinforcer, but the opposite relationship when shock was the reinforcer. This experiment seemed to indicate that learning about audiovisual stimuli and taste stimuli depended on the motivational system engaged, and that such stimuli were not simply interchangeable as traditionally believed. Learning about poisoned food also appeared to be distinguished from other types of learning in that learning could occur even when long delays intervened between the taste stimulus and the induction of illness (for example, Garcia *et al.* 1966). These and other examples of biological constraints have attracted many investigators, and the resulting research has made substantial contributions to learning theory. As an excellent recent review (Domjan 1983) makes clear, however, research on biological constraints has not forced psychologists to abandon general process theories. Rather, proposals have been made for the accommodation of the findings within general process theories, and new experiments have been conducted, finding parallel results in tasks other than, say, poison-aversion learning. It has been proposed (for example, Testa & Ternes 1977) that one factor favouring the selective association of taste cues with poisoning might be similarity (in, for example, temporal pattern) between the conditional and the unconditional stimulus (c.s. and uc.s.). As indirect support for this view, there is now good evidence that similarity of c.s. and uc.s. does facilitate conditioning in tasks not involving poison aversion (for example, Testa 1975; Rescorla & Furrow 1977). Similarly, it has been suggested that long-delay learning in poison avoidance studies may occur because taste stimuli encounter few potentially interfering stimuli in the time intervening before uc.s. arrival (Revusky 1971). This proposal led to food-reinforced maze learning studies in which procedures were introduced to minimize intradelay sources of interference, and learning was indeed demonstrated despite delays of up to 1 h between choices and rewards (Lett 1975). Now it is far from proved that those two factors (c.s.–uc.s. similarity and interference-reduction) provide complete (or even partial) explanations of either selective association formation or long-delay learning in the poison-aversion paradigm. On the other hand, it is equally unclear (Domjan 1983) that there remain any examples of biological constraints that cannot be accommodated within general-process theory. Rather than posit special learning subsystems for each behavioural category, it would seem parsimonious at present to assume the reality of general processes of learning, applicable across many tasks.

Before leaving this topic, it should be added that discussion of it has been brief, partly because reports of 'biological constraints', while suggesting that there might be unexpected specificities between cues and reinforcers, did not in any case necessarily threaten the notion of general intelligence. It might well be the case that there exist some reinforcer-specific adaptations, knowledge of which would be a necessary adjunct to any general process theory, so that certain limitations to that theory's applicability could be understood. It would then be necessary to compile a catalogue of such constraints, but sensible, nevertheless, to continue the attempt to formulate a general theory. So that, while it is argued here that there is as yet no convincing demonstration of a selectivity in association formation which is not the consequence of a general principle applicable to all reinforcers, such a demonstration would not necessarily discredit the attempt to compare intelligence across species.

(*b*) *Optimal foraging strategies*

A further proposed type of reinforcer-specificity may pose a more systematic threat to general process theory since it suggests that different types of reinforcer engage different behavioural problem solving strategies. The proposal in question derives from optimal foraging theory. That theory proposes that the foraging behaviour of animal species has evolved so as to optimize the ratio of energy intake to energy expenditure, and generates predictions about, for example, prey or food type selection, and search patterns both between and within localities. Consideration of this theory inevitably invites speculation on the nature of the mechanisms that might be used in achieving optimal performance. For search patterns, two clear possibilities emerge. On the one hand, a species may assess the probability of food (for example) being in one place rather than another by using general mechanisms, used in the assessment of probabilities of co-occurrence of stimuli, responses and events of all kinds. On the other hand, individual species may have evolved foraging stragegies adapted to the specific type of distribution of the different resources within their niches, a particular strategy being engaged when foraging for one type of resource (for example, food) but not for another (for example, water) which might have a quite different distribution. This latter possibility would have important implications for the majority of investigations of intelligence carried out by comparative psychologists. For results of studies of problem solving motivated by food reward would be largely irrelevant to problem solving motivated by water reward, and so would not throw light on mechanisms of general intelligence (supposing that such mechanisms indeed exist independent of resource-specific strategies).

If there are foraging-specific learning mechanisms, then it should be possible to find species whose distribution of some natural resource differs systematically, and to show differences in their foraging behaviour related to the difference in resource distribution. Moreover it seems likely that such differences should be detectable in laboratory-based experiments as well as in natural environments. Given that animals do solve problems in artificial environments, it seems plausible that, if specific learning mechanisms are involved, then food-rewarded problems would be solved by engaging mechanisms adapted to foraging for food. Not only is it plausible, there is indeed evidence (for example, Lea 1979) that the pattern of both successful and unsuccessful predictions derived from optimal foraging theory found in natural environments is duplicated in laboratory tests.

The radial maze provides, according to some workers, an example of a task in which the performance of different species may reflect differences in foraging behaviour, evolved for contrasting types of food distribution. The radial maze consists of a set of arms (usually eight), each of which has a goal box at its end, radiating from a central platform. Each goal box contains a small, limited amount of food so that a food-deprived animal, starting from the central platform, optimizes its intake : effort ratio by entering each of the eight arms (and consuming the food in its goal box) once only in a series of eight choices. Olton & Samuelson (1976) found that rats performed with remarkable efficiency in the radial maze, averaging more than seven choices of different arms in a series of eight choices. Olton has since suggested (for example, Olton & Schlosberg 1978) that one factor contributing to the rats' success is their tendency, having found food in one place, to avoid that place. This tendency may be innate, since it is seen in young (immediately post-weaning) rats (Olton & Schlosberg 1978). A number of authors have followed Olton's suggestion that the 'win-shift' strategy is a foraging-specific adaptation, to be anticipated in animals whose 'food sources are dispersed or require time to be replenished so that obtaining some food in one area decreases the likelihood of finding additional food there' (Olton & Schlosberg 1978, p. 809).

Bond *et al.* (1981) compared the performance of rats and pigeons in a conventional eight-arm radial maze, using food reward. On the assumption that the natural food resources of the pigeon tend to be 'concentrated and dependable', in contrast to the 'diffusely distributed, irregularly available, and readily depleted' (p. 575) resources of the rat, Bond *et al.* predicted that pigeons would show poor 'spatial event memory', this latter prediction reflecting the notion that pigeons need only remember where food is normally available, not the series of places from which food has recently been obtained. However remarkable it may have been to suppose that homing pigeons, legendary navigators, should show poor spatial event memory, the results were in agreement with their predictions: pigeons made significantly more errors (re-selection of arms already visited) than rats. There is, however, good cause to doubt both the proposal that the difference observed reflects a difference in cognitive capacity and the explanation given for the difference.

The first response to the finding of Bond *et al.* is to ask whether the difference in performance might not have been due to contextual variables, rather than to intellectual differences. Mizumori *et al.* (1982), for example, reported a failure by mice to show spatial memory in a radial maze. The dimensions of the maze were smaller than those normally used, but juvenile rats tested in the identical maze showed efficient spatial memory. It might have been supposed, therefore, that the spatial memory of mice was inferior to that of rats. A recent report by Pico & Davis (1984) suggests, however, that despite the superficial comparability of the rat and mouse visual systems, the species difference in performance was largely due to perceptual factors: mice showed good spatial memory when the arms of the maze were made visually more discriminable. There is evidence also that the performance of rats in eight-arm radial mazes may differ significantly according to physical properties of the maze (for example, Markowska *et al.* 1983). There may, therefore, have been some property of the maze used by Bond *et al.* that made it less well adapted to pigeons than rats. Indeed, some support for this proposal is provided by the observation by Bond *et al.* that the pigeons showed little spontaneous exploration of their maze, so that on occasion the experimenters had to drive pigeons from an explored arm back to the central choice area.

Grounds for supposing that pigeons may be capable of better performance than that reported by Bond *et al.* are provided by a study by Wilkie *et al.* (1981), which used ring doves (*Streptopelia risoria*), birds that are closely related to pigeons (both species belong to the same order, Columbiformes) and that rely on similar food resources. The birds were tested in a 14-arm maze, which consisted of two banks of seven tubes, one above the other, radiating out from a central well. There was a perch at the entrance of each arm, and the doves flew from perch to perch, or from perch to floor or perch. Four birds were allowed to make 14 choices each day, and at asymptote the mean number of correct choices was approximately 11. There was considerable variation from day to day, and from bird to bird, one dove showing 13 correct choices (out of 14) on six of the final 20 sessions of testing. That same bird showed, over those final 20 sessions, no errors over the first six choices of each session, and this contrasts with the sharp fall in accuracy seen on the pigeons' sixth choices (a mean of approximately 60% correct) in the Bond *et al.* report. Wilkie & Summers (1982) note that the doves in the apparatus used by Wilkie *et al.* did not, unlike the pigeons of the Bond *et al.* study, show marked tendencies to visit adjacent arms, and suggest that this may have contributed to their superior performance. We may therefore note that ring-doves are capable of remarkably accurate performance in a 14-arm radial maze and reserve judgement on the question whether their relatives, the pigeons, which appear to face similar foraging demands, might not also possess a capacity superior to that suggested by the Bond *et al.* report. It appears unlikely, to say the least, that a real difference in radial maze capacity between the species could in any case be accounted for in terms of contrasts in distribution of food resources.

A further issue of relevance here is whether pigeons, as might be expected from the 'resource-distribution hypothesis' (Bond *et al.* 1981), in fact show a preference for win-stay as opposed to win-shift strategies in food-rewarded tasks. A clear answer to this question has been provided by Olson & Maki (1983), who explored pigeons' delayed alternation performance in a T-maze. Each trial consisted, first, of a rewarded forced choice to one or other goal box, followed, after a delay, by a free choice in which only the box not entered on the forced choice contained food. The pigeons rapidly mastered this 'win-shift', task, and also performed accurately despite long delays (up to 16 min) between the forced and the free choice. This latter finding indicates that, just as the 'spatial event' memory of the ring dove enjoys a large capacity in terms of the number of items that may be retained, the spatial event memory of the pigeon enjoys impressive capacity in terms of the duration for which information may be retained. Olson & Maki went on to test birds in a 'win-stay' version of the same task, in which the goal box rewarded in the free choice was the same as that entered in the forced choice; the pigeons performed very poorly in this task, averaging fewer than 50% correct choices throughout the period of testing, and this was true whether or not the goal-box food in the forced choice was wholly or only partly consumed by the pigeon. It appears that pigeons, like rats, have a strong tendency to adopt win-shift as opposed to win-stay strategies in certain food-rewarded tasks.

There is evidence (Roitblat *et al.* 1982) that Siamese fighting fish (*Betta splendens*) also have a predisposition to adopt win-shift, as opposed to win-stay strategies, and the evidently widespread occurrence of the strategy in vertebrates clearly raises doubts concerning the resource-distribution account of its origin. Gaffan & Davies (1981, 1982) go further, and argue that the label 'win-shift' is misleading, since it is not, it appears, reward that induces shifting in rats: rats, having explored one place tend to explore an alternative location, whether or not reward has been obtained. Gaffan & Davies (1982) show, moreover, that this tendency simply to shift from a location is in fact weakened by the obtaining of a reward in that place. It seems that we are seeing the operation of a tendency spontaneously to shift, which may reflect some information-gathering exploratory tendency. Such tendencies are seen not only with respect to places, but to other stimuli also: monkeys, for example, having been shown a novel object, show a strong preference, when that object is subsequently re-presented along with a new object, to select the new object, whether or not the original presentation had been rewarded (for example, Mishkin *et al.* 1962). The tendency to shift is not so strong, at least in rats, as to be ubiquitous: rats do not, for example, show spontaneous alternation in jumping stands, or in mazes in which they have to jump a gap to move from arm to arm (Jackson 1941). The tendency need not prevent the development of highly efficient win-stay performance: monkeys performing learning set tasks, despite an initial tendency to show 'win-shift' behaviour, do, as is well known, become highly proficient performers in that task, which, of course, requires adoption of a win-stay, lose-shift strategy with respect to objects.

We have considered cases in which different species have been contrasted according to differences in the distribution of their food supplies. Equally relevant is the search for contrasts within a species between behaviours motivated by types of reward having different 'natural' distributions. A number of authors have, for example, suggested that water is a resource that should, at least in the case of rats, be more dependably found repeatedly in the same place than food. Perhaps the earliest claim to have detected differences in learning that might reflect such a difference was made by Petrinovich & Bolles (1954), who reported that hungry rats rewarded for alternating choices in a T-maze performed more efficiently than thirsty rats in the same task, and suggested that the 'stereotypy' of thirst-motivated behaviour reflected the dependability of water resources. A subsequent report (Bolles & Petrinovich 1956) came, however, to a very different conclusion. The deprivation techniques used in the original report had resulted in loss of mass over the course of the experiment by the hungry rats, but in a gain in mass by the thirsty rats. The second study introduced deprivation techniques which allowed thirsty rats to lose mass, and hungry rats to gain mass, and showed that, irrespective of the type of motivation, rats that lost mass showed efficient alternation performance, while those that gained mass did not.

Subsequent experiments have also failed to find any weaker a tendency in thirsty than in hungry rats to adopt 'win-shift' strategies. Kraemer *et al.* (1983) found excellent performance in a radial maze by thirsty rats, and these authors describe an (unpublished) study by R. H. I. Dale and W. A. Roberts, in which rats were run in a radial maze either hungry or thirsty, and at either 100% or 85% of their *ad libitum* feeding masses. At both levels of body mass, the thirsty rats performed more accurately than the hungry rats. Kraemer *et al.* conclude: 'thus, the suspicion that accurate choice performance in the radial maze is dependent upon general behavioural characteristics associated with food foraging may not be entirely accurate' (p. 380). We can only agree: just as there does not seem to be evidence available to show that different species solve problems in which a common type of reinforcer is used in different ways, those ways reflecting contrasts in the distribution of that reinforcer type, there is, similarly, no evidence that a given species solves problems in which different reinforcers are used by engaging different mechanisms of learning.

4. The null hypothesis

The preceding sections have argued that the appeal to contextual variables as the preferred explanation of species differences in performance on learning tasks is justifiable, that there is as yet no convincing demonstration of a species-specific biological constraint on the associability of particular stimuli or reinforcers, and that species-specific foraging strategies do not appear to determine the course of learning in laboratory-based learning tasks. Discussion of those topics was motivated by their being sources of possible objection to the rationale and methodology of comparative psychologists' investigations of intelligence. If those objections are to be rejected, attention must return to the search for an explanation of the lack of demonstrations of differences in intelligence among vertebrate species.

The position advocated here is based on the proposition that the null hypothesis tested in comparative investigations of any behavioural trait must be that there are no differences among species in that trait. In the absence of experimental proof of any difference the scientist's conclusion should therefore be that there are no differences. In the present case we should, then, conclude that there are no differences, either qualitative or quantitative, among vertebrates (excluding man). In common with all scientific hypotheses this null hypotheses cannot be proved, only disproved; support for the hypothesis will grow as the number of failures to disprove it increases, and to the extent that those failures represent convincing disproofs of specific alternative hypotheses. There is, however, one interesting corollary of the hypothesis that merits particular attention, and that is, that the intellectual demands of any task that is solved by one (non-human) vertebrate may be met by any other vertebrate. So, all the intellectual achievements of chimpanzees should be, contextual variables apart, within the compass of other vertebrates, including, of course, non-mammals. One difficulty with this proposal is that many of the most impressive achievements of the chimpanzee have involved the use of its limbs, and this poses for comparative psychologists the challenge of devising for vertebrates without comparable limbs, tasks whose formal intellectual demands parallel those made in the tasks mastered by chimpanzees. Two recent examples of the use of such tasks, both involving avian subjects, will be discussed here.

The first is taken from a series of studies by Epstein and his colleagues (known as the Columban Simulation Project), each of which attempts to demonstrate in pigeons performance formally comparable to that seen in chimpanzees or other 'higher' vertebrates, performance widely assumed to reflect the superior intellectual capacity of those species. The experiment to be described is an investigation of 'insight' and was modelled on the famous report of Kohler (1925), which has been cited widely as a demonstration of the superior intellect of chimpanzees. In Kohler's original report, a banana was suspended from the ceiling of a room out of reach of a group of six chimpanzees. A wooden crate was available in the room, some distance from the banana. After vain efforts to leap for the banana, and after much pacing up and down, one of the chimpanzees, Sultan, suddenly moved the crate beneath the banana, climbed on it, and now successfully leaped to seize the banana. The novelty and abruptness of the solution convinced Kohler that a process of insight in the chimpanzee had been demonstrated.

Birch (1945) has shown that other apparently insightful solutions are not obtained from chimpanzees without previous manipulatory experience of the objects involved in the solution, and it is reasonable to suppose that Kohler's chimpanzees had indeed previously both moved and stood on boxes. Epstein *et al.* (1984) therefore carried out an experiment, the first phase of which involved training pigeons to move and to stand on, a small cardboard box. In some sessions, the pigeons were trained to move a cardboard box towards a green spot placed at some point at the foot of one of the walls of the apparatus (pushes in the appropriate direction being rewarded with food). In other sessions, pigeons were trained to peck a small toy banana suspended from the ceiling of the apparatus; the banana could be reached by standing on the cardboard box, now placed (by the experimenters) directly beneath it. The box and banana were in different locations in different sessions, there was no green spot, and pushing the box was not rewarded; flying up to peck the banana was also not rewarded, and so extinguished. These two habits, directional pushing, and standing on the box, were established after considerable training, and then a test session was carried out, in which the banana was suspended from the ceiling, the box was present in the chamber at some distance from the banana, and no green spot was presented. The pigeons behaved in essentially the same way as Sultan: they showed initial vain efforts to stretch up to reach the banana, restless pacing, and then 'each subject began rather suddenly to push the box in what was clearly the direction of the banana.... Each subject stopped pushing in the appropriate place, climbed, and pecked the banana' (Epstein *et al.* 1984, p. 61).

We are not concerned here to discuss the nature of the solutions of such problems, whether insight, for example, is a necessary or a useful concept, the key point of interest is the parallel between the chimpanzee and the pigeon performance. There is clearly every reason to suppose that the pigeons solved the problem in exactly the same way as the chimpanzee.

From one of the earliest examples of the supposed intellectual pre-eminence of apes, we may turn to a contemporary example, the series of experiments (for example, Gardner & Gardner 1969; Patterson 1978; Terrace *et al.* 1979) on teaching sign language to apes. Early attempts to teach chimpanzees to speak (for example, Hayes 1961; Kellogg 1968) found rather poor language acquisition, with production of very few words by the subjects. There are, however, reasons for supposing that the physical characteristics of the chimpanzee vocal tract render it incapable of producing the whole range of sounds used in human speech. Recent experiments have, therefore, attempted to teach language without involving speech, by using either American Sign Language, designed for the use of deaf humans (see, for example, Gardner & Gardner 1969; Patterson 1978) or wholly artificial languages (see Premack 1971). These investigators have been markedly more successful than the original speech-based studies, and have, for example, reported successful acquisition of vocabularies of more than 100 words (see Patterson 1978). Now if all vertebrates are capable of comparable intellectual attainment, it should be possible to carry out language-acquisition experiments by using species other than apes, with a comparable degree of success. Birds are clearly not suitable subjects for experiments on sign-language acquisition, but, equally clearly, in birds capable of imitating human speech we have subjects eminently suitable (as apes are not) for speech-based language-acquisition work. The question then is, are birds such as parrots or mynah birds capable of a comparable degree of language acquisition by using speech as are apes, who use sign language?

Mowrer (1950) attempted to train a number of birds of different species (including parrots, a mynah bird and parakeets) to talk, but his training procedures were not particularly systematic, and do not appear likely to have endowed sounds with meaning. Mowrer began by following the utterance (by the experimenter) of a given word with food; when this procedure succeeded in obtaining production of the same word by the bird, the bird was rewarded, again with food. In essence, the birds were taught to imitate sounds produced by their trainers, those sounds having no referent for the birds (except, perhaps, food). Mowrer did succeed in eliciting some words from his birds, but his results were unimpressive, and the method used allows little confidence in Mowrer's claim that birds seem wholly incapable of sentence-formation.

Recent results of an investigation with an African grey parrot (*Psittacus erithacus*) have been more encouraging (Pepperberg 1981, 1983). The parrot (Alex) has been trained by using a technique whereby he is shown an object of interest to him (not a food object) which he is given only if he produces its name; its name he learns originally from dialogue between two trainers. Although Alex has as yet a relatively small vocabulary, consisting of five colour adjectives, four phrases describing shape, and nine nouns, he can combine them appropriately to refer to more than 50 different objects (Pepperberg 1983). There is evidence also that Alex has not learned simply what sound or series of sounds is associated with what object. Alex can respond appropriately to the questions: 'what colour?' and 'what shape?', shown an object for which he has names for both its colour and shape (Pepperberg 1983). This is true for novel objects, and provides evidence for the possession by the parrot of relatively abstract concepts.

Pepperberg's parrot has not yet equalled the achievements of the various apes trained in sign language: his vocabulary, for example, is much smaller than those of the apes. But as the single avian subject yet exposed to an appropriate training schedule, he gives good support to the view that the parrot's talent for language acquisition may not be significantly different from the ape's.

5. Human intelligence

Introduction of experiments on language acquisition in birds and apes leads finally to a brief discussion of the issue which is, perhaps, of the most general significance, namely, the status of human intelligence. While it has been argued here that there are no differences among the intellects of non-human vertebrates, it was acknowledged that humans outstrip other vertebrates in problem-solving capacity. I have argued elsewhere (Macphail 1982) that this capacity cannot be divorced experimentally from the capacity for language, and that the essence of the problem is this: do humans acquire language because they are more intelligent than non-humans, or are humans more intelligent than non-humans because they do acquire language? Of direct relevance to this question is work on language acquisition in non-humans: if non-humans can acquire language, albeit less efficiently than humans, then perhaps the safest conclusion would be that humans acquire language more efficiently because they are (quantitatively) more intelligent. A review of relevant studies (Macphail 1982) concluded, first, that sentence production was the essential characteristic of human language, and, second, that apes do not produce sentences. The most economical account of the data appears, therefore, to be that humans possess a species-specific language-acquisition device, and that it is this qualitative difference alone that distinguishes their intelligence from that of non-human vertebrates.

Scientific hypotheses are commonly, and rightly, judged by two standards, those of parsimony and plausibility. The position outlined above has been argued throughout in terms of parsimony, and it may be thought that its principal weakness is found when it is judged by the criterion of plausibility. It seems likely that the plausibility of these notions as applied to non-humans will grow as demonstrations accrue of hitherto unsuspected capacities in non-mammals (the experiments of Epstein and Pepperberg serve as examples).

There are, however, consequences of this parsimonious account of human intelligence that may be more difficult to accept. For example, humans without language would, according to this view, be no more intelligent than non-human vertebrates. In other words, an otherwise normal human not exposed to language and not educated in a language-using society would be no more intelligent than a chimpanzee (and so, no more intelligent than a goldfish). This may seem extreme, but does invite interesting questions. What problems could the language-deprived human solve that the chimpanzee could not? How would a human intellect locked within a goldfish's body demonstrate its powers? Reflection on these questions may suggest novel ways of seeking to reject the null hypothesis as applied to non-humans, and efforts to reject that hypothesis, whether or not they succeed, are surely the best way to expose the nature of animal intelligence.

I am grateful to Peter Bailey, Geoff Hall, Steve Reilly and Peter Thompson for their helpful comments on an earlier draft of this paper, and to the U.K. Medical Research Council for their support.

References

Birch, H. G. 1945 The relation of previous experience to insightful problem solving. *J. comp. Psychol.* **38**, 367–383.
Bitterman, M. E. 1965 The evolution of intelligence. *Scient. Am.* **212**, 92–100.
Bolles, R. & Petrinovich, L. 1956 Body weight changes and behavioral attributes. *J. comp. Physiol. Psychol.* **49**, 177–180.
Bond, A. B., Cook, R. G. & Lamb, M. R. 1981 Spatial memory and the performance of rats and pigeons in the radial-arm maze. *Anim. Learn. Behav.* **9**, 575–580.
Dickinson, A. 1980 *Contemporary animal learning theory*. Cambridge: University Press.
Domjan, M. 1983 Biological constraints on instrumental and classical conditioning: implications for general process theory. In *The psychology of learning and motivation* (ed. G. H. Bower), vol. 17, pp. 215–277. New York: Academic Press.
Engelhardt, F., Woodard, W. T. & Bitterman, M. E. 1973 Discrimination reversal in the goldfish as a function of training conditions. *J. comp. Physiol. Psychol.* **85**, 144–150.
Epstein, R., Kirshnit, C. E., Lanza, R. P. & Rubin, L. C. 1984 'Insight' in the pigeon: antecedents and determinants of an intelligent performance. *Nature, Lond.* **308**, 61–62.
Gaffan, E. A. & Davies, J. 1981 The role of exploration in win-shift and win-stay performance in a radial maze. *Learn. Motiv.* **12**, 282–289.
Gaffan, E. A. & Davies, J. 1982 Reward, novelty and spontaneous alternation. *Q. Jl exp. Psychol.* **34B**, 31–47.
Garcia, J., Ervin, F. R. & Koelling, R. A. 1966 Learning with prolonged delay of reinforcement. *Psychon. Sci.* **5**, 121–122.

Garcia, J. & Koelling, R. A. 1966 Relation of cue to consequence in avoidance learning. *Psychon. Sci.* **4**, 123–124.
Gardner, R. A. & Gardner, B. T. 1969 Teaching sign language to a chimpanzee. *Science, Wash.* **165**, 664–672.
Hayes, C. 1961 *The ape in our house.* New York: Harper.
Jackson, M. M. 1941 Reaction tendencies of the white rat in running and jumping situations. *J. comp. Psychol.* **31**, 255–262.
Jensen, A. R. 1980 *Bias in mental testing.* London: Methuen.
Kellogg, W. N. 1968 Communication and language in the home-raised chimpanzee. *Science, Wash.* **162**, 423–426.
Kohler, W. 1925 *The mentality of apes.* London: Routledge and Kegan Paul.
Kraemer, P. J., Gilbert, M. E. & Innis, N. K. 1983 The influence of cue type and configuration upon radial-maze performance in the rat. *Anim. Learn. Behav.* **11**, 373–380.
Lea, S. E. G. 1979 Foraging and reinforcement schedules in the pigeon: optimal and non-optimal aspects of choice. *Anim. Behav.* **26**, 875–886.
Lett, B. T. 1975 Long delay learning in the T maze. *Learn. Motiv.* **6**, 80–90.
Mackintosh, N. J. 1970 Attention and probability learning. In *Attention: contemporary theory and analysis* (ed. D. I. Mostofsky), pp. 173–191. New York: Appleton-Century-Crofts.
Mackintosh, N. J. & Cauty, A. 1971 Spatial reversal learning in rats, pigeons, and goldfish. *Psychon. Sci.* **22**, 281–282.
Macphail, E. M. 1982 *Brain and intelligence in vertebrates.* Oxford: Clarendon Press.
Macphail, E. M. & Reilly, S. 1983 Probability learning in pigeons (*Columba livia*) is not impaired by hyperstriatal lesions. *Physiol. Behav.* **31**, 279–284.
Markowska, A., Buresova, O. & Bures, J. 1983 An attempt to account for controversial estimates of working memory persistence in the radial maze. *Behav. Neural. Biol.* **38**, 97–112.
Mishkin, M., Prockop, E. S. & Rosvold, H. E. 1962 One-trial object-discrimination learning in monkeys with frontal lesions. *J. comp. Physiol. Psychol.* **55**, 178–181.
Mizumori, S. J. Y., Rosenzweig, M. R. & Kermisch, M. G. 1982 Failure of mice to demonstrate spatial memory in the radial maze. *Behav. Neural. Biol.* **35**, 33–45.
Mowrer, O. H. 1950 *Learning theory and personality dynamics.* New York: Ronald Press.
Olson, D. J. & Maki, W. S. 1983 Characteristics of spatial memory in pigeons. *J. exp. Psychol: Anim. Behav. Processes* **9**, 266–280.
Olton, D. S. & Samuelson, R. J. 1976 Remembrance of places past: Spatial memory in rats. *J. exp. Psychol: Anim. Behav. Processes* **2**, 97–116.
Olton, D. S. & Schlosberg, P. 1978 Food searching strategies in young rats: win-shift predominates over win-stay. *J. comp. Physiol. Psychol.* **92**, 609–618.
Patterson, F. G. 1978 The gestures of a gorilla: language acquisition in another pongid. *Brain Lang.* **5**, 72–97.
Pepperberg, I. M. 1981 Functional vocalizations by an African Grey parrot (*Psittacus erithacus*). *Z. Tierpsychol.* **55**, 139–160.
Pepperberg, I. M. 1983 Cognition in the African Grey parrot: preliminary evidence for auditory/vocal comprehension of the class concept. *Anim. Learn. Behav.* **11**, 179–185.
Petrinovich, L. & Bolles, R. 1954 Deprivation states and behavioral attributes. *J. comp. Physiol. Psychol.* **47**, 450–453.
Pico, R. M. & Davis, J. L. 1984 The radial maze performance of mice: assessing the dimensional requirements for serial order memory in animals. *Behav. Neural. Biol.* **40**, 5–26.
Premack, D. 1971 Language in chimpanzees? *Science, Wash.* **172**, 808–822.
Rescorla, R. A. & Furrow, D. R. 1977 Stimulus similarity as a determinant of Pavlovian conditioning. *J. exp. Psychol: Anim. Behav. Processes* **3**, 203–215.
Revusky, S. H. 1971 The role of interference in association over a delay. In *Animal memory* (ed. W. K. Honig & P. H. R. James), pp. 155–213. New York: Academic Press.
Roitblat, H. L., Tham, W. & Golub, L. 1982 Performance of *Betta splendens* in a radial arm maze. *Anim. Learn. Behav.* **10**, 108–114.
Terrace, H. S., Pettito, L. A., Sanders, R. J. & Bever, T. G. 1979 Can an ape create a sentence? *Science, Wash.* **206**, 891–902.
Testa, T. J. 1975 Effects of similarity of location and temporal intensity pattern of conditioned and unconditioned stimuli on the acquisition of conditioned suppression in rats. *J. exp. Psychol: Anim. Behav. Processes* **1**, 114–121.
Testa, T. J. & Ternes, J. W. 1977 Specificity of conditioning mechanisms in the modification of food preferences. In *Learning mechanisms in food selection* (ed. L. M. Barker, M. R. Best & M. Domjan), pp. 229–253. Waco, Texas: Baylor University Press.
Wilkie, D. M., Spetch, M. L. & Chew, L. 1981 The ring dove's short-term memory capacity for spatial information. *Anim. Behav.* **29**, 639–641.
Wilkie, D. M. & Summers, R. J. 1982 Pigeons' spatial memory: factors affecting delayed matching of key location. *J. exp. analyt. Behav.* **37**, 45–56.

Discussion

H. B. Barlow, F.R.S. (*University of Cambridge, U.K.*). There is another hypothesis that has already been alluded to by the previous speakers, and I would like to ask whether you think it has been excluded by the comparative learning studies you have reviewed. The hypothesis is that intelligence varies greatly between species, but is concerned primarily with the formation of a working model of the normal environment that an animal lives in. Do the learning tests you review exclude the possibility that the accuracy and completeness of this model varies greatly between species? And if the answer is 'no', might it not be a more fruitful hypothesis than your third – null – hypothesis?

Let me point out that if intelligence depends upon the accuracy and completeness of the cognitive model, this does not reduce the importance of language for intelligence, but rather offers an explanation for why it is so important: language makes it possible to share and spread the model among many members of a community, and (especially for written language) it allows this model to persist in time and to be successively improved. As Jerison said, a human utterance can communicate part of the speaker's model.

On this view, your null hypothesis may possibly hold for learning ability as tested by experimental psychologists, but does not hold for intelligence.

E. M. Macphail. Although it is difficult to see how differences in cognitive models might be measured, there can be no doubt that animals of different species do form very different models of their environments, and no doubt that these models play a role in the general adaptability of animals. But an important factor in the formation of such models must be the quality of sensory information available, and there is equally no doubt that sensory capacities vary widely. Unless it can be demonstrated that species differences in cognitive models are not a consequence of differences in sensory capacities, I see no need to accept the argument that differences in models reflect (or constitute) differences in intelligence. To put the case in a somewhat crude way: the environmental models formed by the blind must be very different from those formed by sighted individuals, and the range of problems soluble by a blind man is indeed much restricted compared with the range soluble by sighted humans, but we would not wish to argue in consequence that the blind are less intelligent than those with sight.

Phil. Trans. R. Soc. Lond. B **308**, 53–65 (1985) [53]
Printed in Great Britain

Differences in mechanisms of intelligence among vertebrates

By N. J. Mackintosh[1], B. Wilson and R. A. Boakes[2]

[1] *Department of Experimental Psychology, University of Cambridge, Downing Street, Cambridge CB2 3EB, U.K.*

[2] *Laboratory of Experimental Psychology, University of Sussex, Brighton BN1 9QG, U.K.*

Much comparative research aimed at establishing differences in intelligence among vertebrates has failed to convince the sceptic, because it has concentrated on a single experimental paradigm (such as learning sets), while employing a diverse array of species in the hope of establishing a rank ordering of intelligence. The sceptic can insist that such research has not even established that there are any differences in mechanisms of intelligence between any pair of vertebrate species, let alone elucidated the nature of the difference, and even the unsceptical will doubt that such research is ever likely to establish a rank order of intelligence.

It is more informative to concentrate on fewer species, but a broader range of experimental paradigms. Thus, studies of serial reversal learning have consistently suggested that goldfish do not show such rapid improvement as do rats. One explanation of this might be that rats learn more effectively than goldfish to use the outcome of one trial to predict the outcome of the next. The suggestion is supported by finding other experimental paradigms, such as alternation learning, which must also tap such a process, and where rats again learn more rapidly than goldfish.

Efficient learning-set performance may also depend on this process, but must in addition require the subject to transfer this rule across changes of stimuli. There is reason to believe that not all vertebrates are equally adept at such transfer, and this possibility is explored in a series of experiments studying the transfer of matching-to-sample in pigeons and corvids. The corvids display significantly better transfer and the close similarity in training procedures possible with these subjects makes it unlikely that this is due to differences in these training procedures.

1. Introduction

Comparative psychologists have spent about 100 years in the more or less serious scientific study of animal intelligence. It cannot be claimed that they have very much to show for their pains. They have discovered a great deal about the processes underlying conditioning and simple associative learning, and have established that some of these processes are of great generality, to be found in most vertebrates and almost certainly in numerous invertebrates also. But there is a persistent belief that simple conditioning is not the only form that animal intelligence can take, and that intelligence should not be so widely and uniformly distributed. If these beliefs are justified, then there is rather little else that animal psychology can point to in the way of solid achievement. Macphail (1982), for example, is able to conclude a painstaking analysis of comparative studies of intelligence with the suggestion that, for all we know of the matter, there is no good reason to assume that there are any differences, either quantitative or qualitative, in the mechanisms of intelligence of non-human vertebrates. Have 100 years of research really produced no evidence one way or other to answer this question? And if not, why not?

In partial defence it is important to acknowledge that there are formidable difficulties in the way of good comparative research. But some have been of psychologists' own making. The history of research on learning sets provides a good example. Harlow (1949) had discovered that although, if trained on a simple discrimination problem (where choice of one alternative is rewarded and choice of the other is not), rhesus monkeys might learn to choose the correct alternative no faster than rats, if they were trained on a whole series of such problems, in each of which a new pair of stimuli was used, the monkeys would show a remarkable improvement and eventually learn each new problem in a single trial. In no time at all, there were studies of learning-set formation in primates ranging from human children to marmosets, and in other mammals from cats to squirrels to rats, and a rank-order of mammalian intelligence, based on this one task, was constructed (for example, Warren 1965; Passingham 1982).

It is quite possible that the differences in performance observed in these learning-set experiments do indeed reflect some differences in mechanisms of learning or intelligence. But such data alone will never be sufficient to establish this conclusion, nor will it greatly help to add yet more species to the list, or to correlate differences in learning-set performance with differences in some notional phyletic status or relative brain size. What is needed is first a realistic acknowledgement of alternative explanations, for example the trivially uninteresting possibility that primates are likely to learn a series of visual discrimination problems more rapidly than rats because they have a more elaborate visual system. But, above all, we need a theory of the nature of these putative differences in learning mechanisms, and we need to test that theory by examining its predictions in other experimental situations.

2. Serial reversal

It will be easier to start with a possibly simpler task – one at which rats at any rate are quite proficient. It has long been known that if rats are trained on a series of reversals of a single discrimination problem, where choice of first one alternative and then of the other is rewarded, the speed at which they learn each new reversal will rapidly increase. Some such improvement is probably the norm in vertebrates: although it was once thought that some teleost fish were unable to learn later reversals any faster than earlier ones, it is now clear that under some conditions of training they can. But, despite suggestions to the contrary, this hardly proves that there are no differences between, say, rats and goldfish in the processes of reversal learning. Figure 1 provides some illustrative examples of the performance of these two groups on visual (brightness or colour) and spatial reversals. It is apparent that the goldfish are notably less proficient than the rats. What is not so immediately apparent, but should be emphasized, is that figure 1 provides a very conservative estimate of this difference. The data for the goldfish have been selected as the two clearest examples available of improvement over a series of reversals.† Those for the rat are by no means so highly selected: although they show one characteristic feature of the rat's performance, namely that later reversals are learned reliably

† An experiment by Setterington & Bishop (1967) has provided clear evidence of improvement over a series of spatial reversals by another species of teleost fish, African mouth breeders. Although the published data are presented in a manner which makes comparison with the data shown in figure 1 impossible, Dr Setterington has kindly provided his original data, from which it is possible to calculate that the fish averaged 9.1 errors per reversal over reversals 1–10, 9.1 over reversals 11–20, and 8.6 over reversals 21–30. After 70 reversals the average number of errors per reversal had dropped to 6.6. The improvement is significant but not obviously any more rapid than that shown by goldfish in figure 1, and, by rat standards, slow.

faster than the original discrimination, there are numerous experiments in which rats have learned later reversals with only a single error.

Figure 1 suggests, therefore, that there are significant differences in the performance of rats and goldfish on a series of discrimination reversals, but it leaves open the possibility that these differences are due to uninteresting differences in experimental procedure. The fact that rats

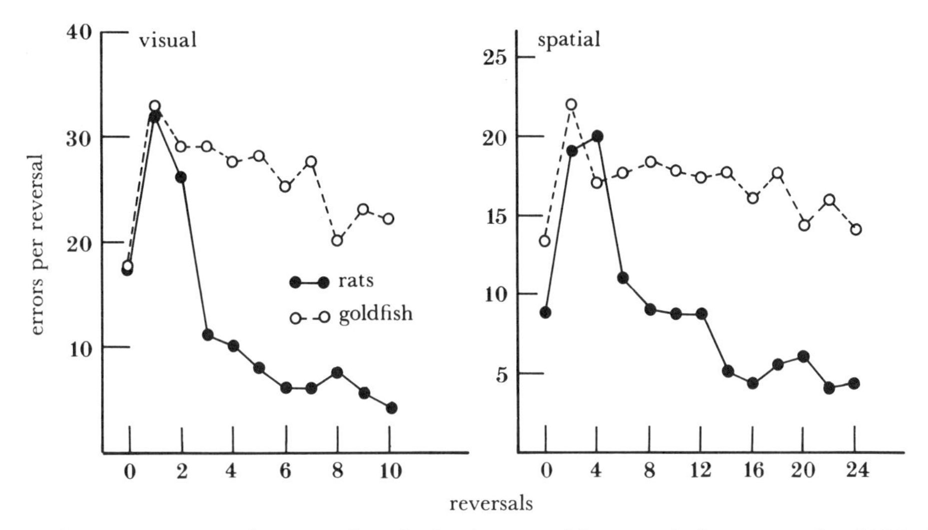

FIGURE 1. Errors per reversal over series of visual or spatial reversals by rats and goldfish (O = original discrimination). The visual reversal data for the rats are from Mackintosh & Holgate (1969) and for the goldfish from Engelhardt *et al.* (1973). The spatial reversal data for both animals are from Mackintosh & Cauty (1971).

learn visual reversals very much faster than goldfish even though taking just as long to learn the original discrimination makes it unlikely that differences in sensory capacity are involved, but speed of reversal is affected by numerous other factors. How could we prove that variations in the contribution of one or more of these factors were not responsible for the difference in performance? It is not at all obvious that any further study of reversal learning by rats or goldfish could ever do this. For, as a matter of fact, just as there are conditions of training under which goldfish show no improvement over a series of reversals, so are there for rats (for example if they are trained at the rate of only one trial per day, Clayton 1962). This hardly proves that the differences shown in figure 1 are artefacts of differences in conditions of training; but it does mean that without some independent criterion for deciding what were functionally comparable conditions, further experiments on reversal learning may never be able to disprove this. And what could that independent criterion be?

Perhaps we should try a different approach. If the difference in reversal performance is due to a difference in learning mechanism, what might this be? One simple possibility is that the highly efficient reversal performance of rats reflects their ability to use a set of cues whose relation to reward stays constant throughout the experiment. The difficulty in reversal learning is that now one stimulus, now the other, predicts reward, and reliance on the physical attributes of these stimuli to predict the outcome of a trial means that each new reversal requires the unlearning of an old predictive relationship and the learning of a new one. But, with the exception of the first trial of each new reversal, there is a consistent relationship between the outcome of one trial and the animal's choice on the preceding trial and the outcome of that choice. On any given trial (but the first), of the two alternatives A and B, A will be rewarded either if A was chosen on the preceding trial and that choice was rewarded, or if B was chosen and that choice was not rewarded.

If rats become proficient at reversal learning because they learn that choice and outcome of choice on one trial predict the outcome of the next trial, and if goldfish are less efficient at detecting or using this relationship, we should expect to see differences in other experimental paradigms. One task that requires an animal to use the outcome of one trial to predict the outcome of the next is alternation learning. In the simplest case, a response such as pressing a lever or panel is rewarded only on every alternate trial. Rats rapidly detect this pattern and learn to respond on rewarded trials following non-reward, but to slow down or refrain from responding on non-rewarded trials following reward. Goldfish, by contrast, find this an extraordinarily difficult task, as figure 2 shows.

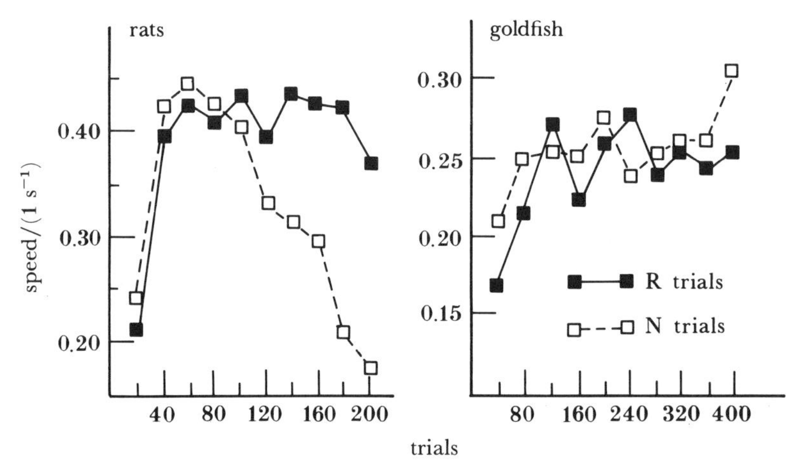

FIGURE 2. Speed of responding on rewarded (R) and non-rewarded (N) trials by rats and goldfish in experiments on alternation learning. The rat data are from Bloom & Malone (1968) and the goldfish data from Mackintosh (1971).

3. TRANSFER OF REVERSAL LEARNING

There are several other experimental paradigms where differences between the behaviour of rats and goldfish are readily understood as differences in the extent to which they detect these sorts of regularities (Mackintosh 1974; Macphail 1982). It does not seem too fanciful to suggest that the ability to track sequential variations and dependencies in the occurrence of food or other events of consequence might be an important component of intelligence. Certainly, it seems demanded by other tasks beloved of comparative psychologists, such as the discrimination learning-set task. The most plausible account of the experienced rhesus monkey's one-trial learning of any new discrimination problem is that he has learned the consistent predictive relationship between choice and outcome on one trial and outcome of the next. But in this case, unlike that of serial reversal, another process would seem to be required. One-trial reversal learning requires that the animal learn that choice of a specific stimulus, A, will be rewarded if, on the previous trial, he either chose A and was rewarded or chose B and was not. In the learning-set task, he must learn that this is true of *any* pair of stimuli, A and B. The rule must be generalized across changes in physical stimuli.

Evidence of such generalization in primates has been obtained from two variants of the serial reversal experiment. In one, the animal is trained on a discrimination between A and B and then on a single reversal of A–B; then on C versus D and its reversal; then on E versus F and so on. Rhesus monkeys, it is well established, rapidly increase their speed of learning new reversals (Warren 1960*a*). Cats have shown some evidence of such generalized improvement at reversal, but without matching the rhesus monkey's ability to reverse any new problem after a single error (Warren 1960*b*). The second variant involves training animals on a series of reversals of a single discrimination and then transferring them to a new problem, for example to a series of new discriminations in the standard learning-set procedure. Chimpanzees and macaque monkeys show excellent transfer under these circumstances (Schusterman 1964; Schrier 1966; Warren 1966). But in the only published study with another mammal, Warren (1966) found no evidence of transfer in cats, even though they had been trained on more than three times as many reversals, to a point where they were making half as many errors per reversal, as monkeys who have shown excellent transfer (Schrier 1966).

An alternative transfer task is learning a series of reversals of a new discrimination. Staddon & Frank (1974) trained pigeons for up to 134 reversals of one discrimination, by using a standard free-operant multiple schedule, before transferring them to a new discrimination and its reversals. Although by the end of training on the original series, all birds consistently made more than 90% of their responses to the correct stimulus on the first day of each new reversal, the change of stimulus markedly disrupted their performance, and it took more than 20 reversals before they were performing at this level of accuracy again. The two birds who were treated identically during their original and transfer reversal series performed no more accurately over the first half dozen reversals of their second problem than they had on the first. Paula Durlach has been undertaking a rather more systematic experiment along these lines. Her preliminary results confirm that pigeons, although showing a progressive reduction in the number of errors per reversal of one discrimination, are disrupted by a change in stimuli. Although there is clear evidence of some transfer from one pair of colours to another, there seems to be little or no transfer if birds are initially trained on reversals of a line orientation discrimination before being transferred to colour reversals. The most plausible interpretation is that the positive transfer in the former case reflects an increase in attention to colour cues, but that there is rather little transfer of any more general process responsible for reversal improvement. This is entirely consistent with the pigeon's relatively inefficient performance on standard learning-set tasks. Wilson (1978) trained pigeons on a series of 1000 visual discrimination problems. By the end of training they were averaging less than 55% correct on trial 2 of each new problem.

The question still remains whether the differences in behaviour suggested by all these experiments imply genuine differences in processes of learning or whether they are uninteresting consequences of some unspecified differences in experimental procedures. The diversity of results we have considered may make this latter alternative seem less plausible, but it cannot easily be ruled out. The experimental arrangements used to study reversal learning in goldfish or learning-set formation in pigeons differ markedly from those used for rats and primates. And even if the apparatus and training procedures were identical, we could not be sure that they were making equivalent demands on, say, the sensory or motivational systems of such diverse animals. It would make more sense to compare the behaviour of animals that did not differ so grossly in bodily structure or phyletic position. Different groups of birds promise to provide suitable material for such comparisons, for Kamil and his colleagues have shown that American blue jays, members of the corvid family, can perform very efficiently on the standard learning-set task, and will show significant transfer from serial reversal of one discrimination to a series of new problems (Kamil *et al.* 1973, 1977).

4. Transfer of matching and oddity in pigeons and corvids

We shall describe a set of experiments studying the transfer of matching and oddity learning in pigeons and European corvids (jackdaws, jays, rooks). Matching and oddity are types of conditional discrimination: animals are required to choose between two 'comparison' stimuli (for example, red versus yellow); which alternative is correct depends on the colour (red or yellow) of the 'sample' stimulus displayed at the outset of the trial. In matching, the correct comparison stimulus is the same as the sample; in oddity the comparison stimulus that is different from the sample is correct. Pigeons and corvids have no difficulty learning matching and oddity discriminations; the question is how they do it, and whether they can transfer the solution across a change of stimuli. Transfer of matching and oddity by pigeons was first studied by Cumming & Berryman (1965); they initially trained pigeons with three colours as stimuli; red, green and blue, and found that when they substituted yellow for blue the birds' performance dropped to chance. Zentall & Hogan (1974, 1975, 1978) adopted a potentially more sensitive, savings measure of transfer. Having trained pigeons on matching or oddity with one pair of stimuli, they transferred them to a new pair, with half the birds continuing on the same kind of problem as before (matching to matching or oddity to oddity) and half reversed to the opposite problem. Zentall and Hogan have consistently reported that the birds continuing on the same kind of problem learn their transfer task faster than those shifted to the opposite problem, and have taken this to mean that pigeons solve matching and oddity discriminations by learning that reward is predicted by the relation (of similarity or difference) between sample and comparison stimulus, and that they can transfer this rule across a change in stimuli.

There are two aspects of Zentall & Hogan's data that should give one slight pause. First, in several of their experiments they have studied transfer between red and green lights in one problem and yellow and blue in the other. But another of their own experiments (Zentall *et al.* 1981) establishes rather clearly that pigeons show substantial sensory generalization between red and yellow and between green and blue lights. The question at issue is certainly not whether transfer can occur when animals treat one pair of stimuli as similar to another. Secondly, in their other studies there was a consistent and striking asymmetry in the transfer found. Although pigeons learned a second discrimination, when it was matching, faster if their first discrimination had also been a matching problem, when the second task was oddity it did not matter whether the first was matching or oddity. Independent replication of their procedures confirms exactly this pattern of results as can be seen in the left-hand panel of figure 3 (Wilson *et al.* 1985*a*). It is only in the case of birds tested on matching, therefore, that there is any reason to postulate transfer of the relational rule. But there is also a reliable asymmetry to the way in which pigeons learn matching and oddity discriminations in the first place, which suggests an alternative explanation even of this apparent transfer. Pigeons have an initial preference for oddity, but may actually take longer to learn oddity to a high level of accuracy than to learn matching (Cumming & Berryman 1965; Wilson 1978). Because their initial preference is so often rewarded when they are trained on oddity, perhaps they learn less and perhaps this explains why they take so long to solve a second problem when it requires something more than responding in accordance with their natural preference. The suggestion is that prior training on oddity produces negative transfer when pigeons are subsequently trained on matching. The suggestion is supported by an experiment (Wilson *et al.* 1985*a*) in which pigeons were trained on a matching discrimination, having previously been trained either on matching or on oddity with a different pair of stimuli, or on a conditional discrimination whose solution did not involve any relation between sample and comparison stimuli (the problem was of the following form: there were two sample stimuli, A or B, and two different comparison stimuli, X and Y. The solution required birds to learn the quite arbitrary rule: if the sample is A, then X is rewarded; if the sample is B, then Y is rewarded). The performance of these three groups of pigeons on transfer to matching is shown in the right-hand panel of figure 3. It is clear that although birds

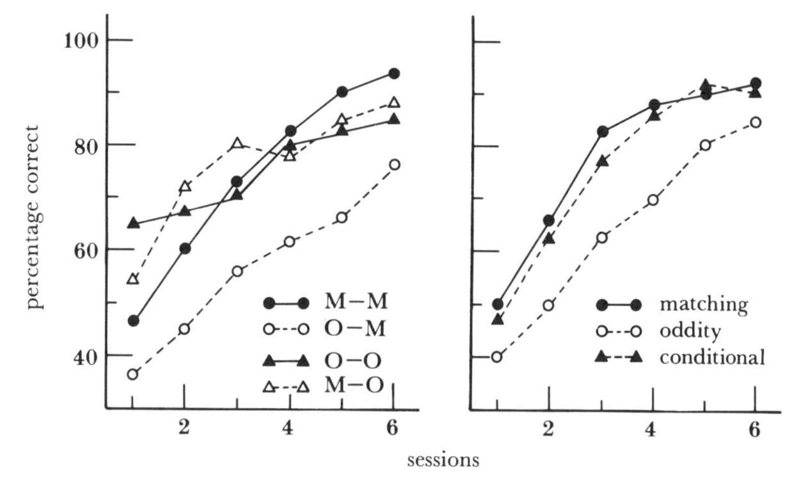

FIGURE 3. Transfer of matching and oddity learning by pigeons. The left panel shows the course of learning matching or oddity to red and yellow stimuli after prior training, on matching or oddity, to blue and green. Group M–M was trained on matching with both sets of stimuli, group O–M on oddity and then transferred to matching, while groups O–O and M–O were transferred to oddity after prior training on oddity or matching respectively. The right panel shows the course of learning matching to red and yellow stimuli after prior training on matching, oddity or conditional discriminations with different colours. Data from Wilson *et al.* (1985*a*).

previously trained on matching learned faster than those previously trained on oddity, they did not learn any faster than those trained on the conditional discrimination. If there is no difference between matching- and conditional-trained birds, the opportunity available to the former, but not to the latter, to learn the matching relation between sample and comparison stimuli cannot have benefited subsequent matching learning. The difference between matching- and oddity-trained birds must have been due to negative transfer from oddity rather than to positive transfer from matching.

There is thus remarkably little evidence for the transfer of matching in pigeons. There are two possible explanations: either pigeons do not, under the conditions of training used so far, solve matching and oddity discriminations in terms of the relation between sample and comparison stimuli; or, although they do learn this relationship, they do not transfer it across a change of stimuli. Three observations suggest that the second alternative is the more likely. First, it is hard to see how the asymmetry between matching and oddity problems alluded to above (for example, the initial preference for oddity) is to be explained without supposing that the pigeon detects the fact that the relation between sample and correct comparison stimulus differs in the two cases. Secondly, there is some evidence that pigeons may, under some circumstances, learn a matching discrimination faster than a non-relational conditional discrimination, suggesting that the presence of the relational cue may facilitate learning (Zentall *et al.* 1984). Thirdly, Zentall *et al.* (1981) have shown that whereas pigeons' performance on a well-trained matching task may be disrupted by changing the correct comparison stimulus, but not by changing the incorrect stimulus, performance on oddity is disrupted by changing the incorrect comparison stimulus, but not by changing the correct one. In each case, performance is disrupted only when the comparison stimulus that is the same as the sample is changed, and the implication must be that this relationship is used in the solution of these problems.

If this is accepted, then the pigeon's apparent failure to transfer matching or oddity to new sets of stimuli is a failure of transfer not of learning. They detect the relationship, but cannot apply it to new stimuli, and their poor performance is consistent with their poor performance on learning sets and the relative lack of transfer following serial reversal training. But how poor is their performance? Is there any reason to believe that any other animal would behave any differently? It is, in fact, well established that both primates and dolphins can solve generalized matching or oddity problems, where the stimuli change on every trial and only the relation between sample and comparison stimuli is available to predict reward (Mishkin *et al.* 1962; Herman & Gordon 1974). But the procedures of these experiments are so different from those used with pigeons that it is difficult to specify the cause of any apparent difference in outcome (see, however, D'Amato & Salmon 1984). What is needed is a comparison of the performance of pigeons and other birds, trained with exactly the same procedures, stimuli and apparatus.

Such a comparison reveals clear and striking differences between pigeons and corvids. Where the pigeons show at best only marginal signs of transfer, evident if at all only from differences in rate of learning, transfer in the corvids is relatively unambiguous and evidenced by significant deviations from chance on the first day of testing (Wilson *et al.* 1985*b*). In one experiment, pigeons and jackdaws were initially trained on either a matching or a non-relational conditional discrimination. Although the procedure was nominally identical for both species, the jackdaws took rather longer to learn these problems than the pigeons, showing for example an even more marked preference for the odd (incorrect) stimulus in the matching problem. In spite of this, when all birds were tested on a new matching problem with a different set of stimuli, it was only the jackdaws that provided evidence of differential transfer. The left panel of figure 4 shows the performance of the pigeons: although learning quite rapidly, the nature of their prior experience had no effect on their performance and both groups scored at chance on the first day. The jackdaws, however, although again learning matching slowly, were affected by their earlier training. Those that had previously learning matching scored above chance on the first day, while those trained on the conditional discrimination scored well below chance (indicating their natural preference for oddity) and the difference between the two groups was maintained for the remainder of the experiment.

To increase the chances that pigeons might show positive transfer, in a second experiment pigeons and European jays were trained on a series of three different matching (or oddity) discriminations, each with a new pair of colours, before a final test problem, where the stimuli were horizontal and vertical lines, on which they were trained either with the same conditions of reward (that is, matching if they had so far learned matching, and oddity if they had so far learned oddity) or reversed. The results, collapsed across matching and oddity, are shown

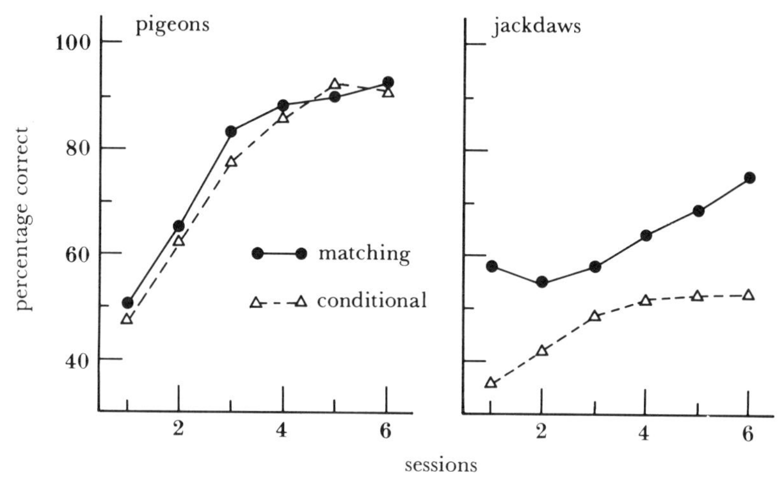

FIGURE 4. Transfer to matching after prior training on either matching or conditional discriminations with different stimuli in pigeons and jackdaws. Data for pigeons from Wilson *et al.* 1985*a* (as shown in figure 3); data for jackdaws from Wilson *et al.* (1985*b*).

in figure 5. Consider first performance on the three colour problems. Although there was no difference between pigeons and jays on the first discrimination, by the time they reached their third problem with the same rule, the jays were learning very much faster than the pigeons, and were performing well above chance on the first day of the problem. On the final problem, both pigeons and jays performed less accurately if they were shifted to the reverse conditions, but whereas this difference was reliable and persisted in jays, in the pigeons it was small, not significant, and was not maintained on subsequent days of testing.

A third experiment used a rather different paradigm for testing transfer across a change in

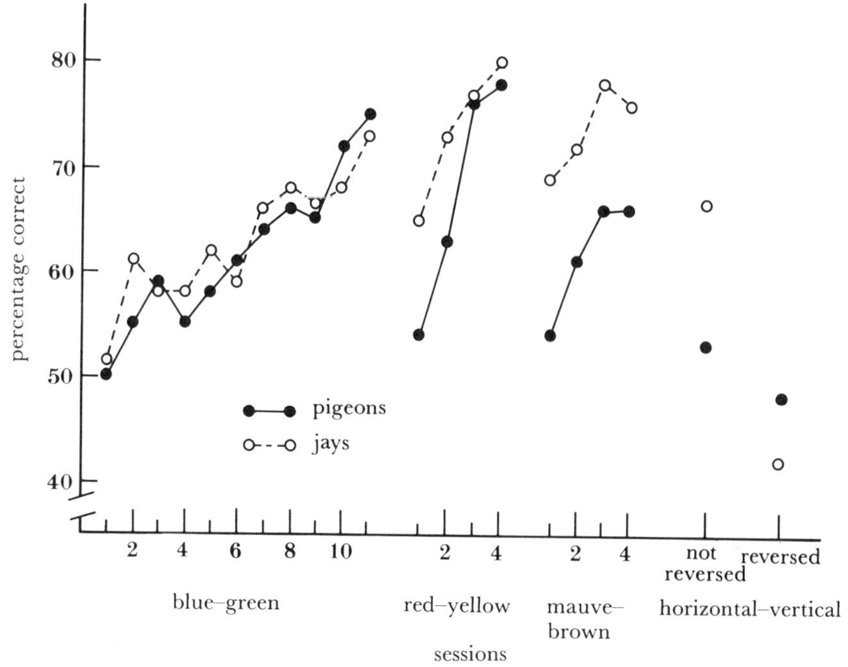

FIGURE 5. Course of learning three consecutive matching or oddity colour discriminations by pigeons and jays, together with transfer to a horizontal-vertical discrimination with reward assignments not reversed or reversed. Data from Wilson *et al.* (1985*b*).

stimuli. Pigeons and rooks were trained to choose between two illuminated panels. If they were the same colour (both blue or both mauve) then a response to the right-hand panel was rewarded; if they were different (one blue and one mauve), a response to the left-hand panel was rewarded. Once they had learned this discrimination (which proved very much harder than ordinary matching or oddity) they were transferred to a new pair of colours, red and yellow, either with the same or with the reversed rule. On the first day of this new problem, rooks scored well above chance (68% correct) if the rule stayed the same, and below chance (40%) if the rule was reversed. Pigeons sccred exactly at chance (50%) in both conditions, and there was no reliable difference between reversed and non-reversed groups over ten sessions of training on the new problem.

The results of these experiments show some rather clear differences between the behaviour of pigeons and corvids trained under comparable conditions. It is always possible to insist that unsuspected differences in the impact of these training procedures on the two groups of birds were in fact responsible for these differences in behaviour. But there comes a point at which this argument starts to lose its force. We know of no plausible candidate for such a critical factor, and if recourse is always to be made to the possibility of such an effect, the position it seeks to defend will become irrefutable in principle, and hence vacuous.

What should be acknowledged, of course, is that these results do not, and could not, prove that pigeons are incapable of transferring a rule across a change in stimuli. There are occasional hints of such transfer in the published data, and it is always possible that other training procedures might succeed more clearly where those tried so far have failed. The history of comparative psychology is replete with categorical, but subsequently recanted, assertions about the limits to a particular animal's capacities. What the present results do suggest is that there is a real difference between pigeons and some other birds (and by extension, therefore, probably a difference between pigeons and, say, some primates) in the extent to which such rules are readily transferred.

5. Conclusions

Our conclusion is that it is possible to identify two processes, use of the outcome of one trial to predict the outcome of another, and transfer of this or other rules across a change in physical stimuli, both of which seem required, for example, for efficient learning-set performance, and neither of which is uniformly distributed among vertebrates. It is possible that differences in these processes may underlie some observed differences in learning-set performance. This is a fairly modest conclusion. We are not advancing a general account of the evolution of intelligence in vertebrates; and we have not even addressed the question why such differences might appear between the rather haphazardly selected animals under discussion. There is little point in considering such larger issues until we have done a great deal more to establish that there are any differences to talk about.

Finally, it is worth stressing that part of our conclusion was not about differences between animals at all, it concerned the identification of certain processes of learning and intelligence. The real interest of the comparison between pigeons and corvids, for example, is not to prove that pigeons are stupid, but rather to take advantage of the pigeon's apparent failure to transfer say, the matching rule to new stimuli, to establish the point that such transfer is not an automatic consequence of teaching an organism a matching discrimination. When transfer does occur, therefore, this is precisely something that demands explanation; it requires the postulation of a process or set of processes that probably lies outside the scope of our simple theories of conditioning or associative learning.

References

Bloom, J. M. & Malone, P. 1968 Single alternation patterning without a trace for blame. *Psychon. Sci.* **11**, 335–336.

Clayton, K. N. 1962 The relative effects of forced reward and forced nonreward during widely spaced successive discrimination reversal. *J. comp. Physiol. Psychol.* **55**, 992–997.

Cumming, W. W. & Berryman, R. 1965 The complex discriminated operant: studies of matching-to-sample and related problems. In *Stimulus generalization* (ed. D. Mostofsky), pp. 284–330. Stanford: University Press.

D'Amato, M. R. & Salmon, D. P. 1984 Cognitive processes in cebus monkeys. In *Animal cognition* (ed. H. L. Roitblat, T. G. Bever & H. S. Terrace), pp. 149–168. Hillsdale, N. J.: Lawrence Erlbaum Associates.

Engelhardt, F., Woodard, W. T. & Bitterman, M. E. 1973 Discrimination reversal in the goldfish as a function of training conditions. *J. comp. Physiol. Psychol.* **85**, 144–150.

Harlow, H. F. 1949 The formation of learning sets. *Psychol. Rev.* **56**, 51–65.

Herman, L. M. & Gordon, J. A. 1974 Auditory delayed matching in the bottlenose dolphin. *J. exp. analyt. Behav.* **21**, 19–29.

Kamil, A. C., Jones, T. B., Pietrewicz, A. & Mauldin, J. E. 1977 Positive transfer from successive reversal training to learning set in blue jays. *J. comp. Physiol. Psychol.* **91**, 79–86.

Kamil, A. C., Lougee, M. & Shulman, R. J. 1973 Learning-set behaviour in the learning-set experienced bluejay. *J. comp. Physiol. Psychol.* **82**, 394–405.

Mackintosh, N. J. 1971 Reward and aftereffects of reward in the learning of goldfish. *J. comp. Physiol. Psychol.* **76**, 225–232.

Mackintosh, N. J. 1974 *The psychology of animal learning.* London: Academic Press.

Mackintosh, N. J. & Cauty, A. 1971 Spatial reversal learning in rats, pigeons and goldfish. *Psychon. Sci.* **22**, 281–282.

Mackintosh, N. J. & Holgate, V. 1969 Serial reversal training and nonreversal shift learning. *J. comp. Physiol. Psychol.* **67**, 89–93.

Macphail, E. M. 1982 *Brain and intelligence in vertebrates.* Oxford: University Press.

Mishkin, M., Prockop, E. S. & Rosvold, H. E. 1962 One-trial object-discrimination learning in monkeys with frontal lesions. *J. comp. Physiol. Psychol.* **55**, 178–181.

Passingham, R. W. 1982 *The human primate.* W. H. Freeman: San Francisco.

Schrier, A. M. 1966 Transfer by macaque monkeys between learning-set and repeated-reversal tasks. *Percept. mot. Skills* **23**, 787–792.

Schusterman, R. J. 1964 Successive discrimination-reversal training and multiple discrimination training in one-trial learning by chimpanzees. *J. comp. Physiol. Psychol.* **58**, 153–156.

Setterington, R. G. & Bishop, H. E. 1967 Habit reversal improvement in the fish. *Psychon. Sci.* **7**, 41–42.

Staddon, J. E. R. & Frank, J. 1974 Mechanisms of discrimination reversal. *Anim. Behav.* **22**, 808–828.

Warren, J. M. 1960*a* Supplementary report: effectiveness of food and nonfood signs in reversal learning by monkeys. *J. exp. Psychol.* **60**, 263–264.

Warren, J. M. 1960*b* Discrimination reversal learning by cats. *J. genet. Psychol.* **97**, 317 327.

Warren, J. M. 1965 Primate learning in comparative perspective. In *Behavior of nonhuman primates: modern research trends* (ed. A. M. Schrier, H. F. Harlow & F. Stollnitz), pp. 249–282. New York: Academic Press.

Warren, J. M. 1966 Reversal learning and the formation of learning sets by cats and rhesus monkeys. *J. comp. Physiol. Psychol.* **61**, 421–428.

Wilson, B. 1978 Complex learning in birds. Unpublished D.Phil. Thesis, University of Sussex.

Wilson, B., Mackintosh, N. J. & Boakes, R. A. 1985*a* Matching and oddity learning in the pigeon: transfer effects and the absence of relational learning. (In preparation.)

Wilson, B., Mackintosh, N. J. & Boakes, R. A. 1985*b* Transfer of relational rules in matching and oddity learning by pigeons and corvids. (In preparation.)

Zentall, T. R., Edwards, C. A., Moore, B. S. & Hogan, D. E. 1981 Identity: the basis for both matching and oddity learning in pigeons. *J. exp. Psychol: Anim. Behav. Proc.* **7**, 70–86.

Zentall, T. R. & Hogan, D. 1974 Abstract concept learning in the pigeon. *J. exp. Psychol.* **102**, 393–398.

Zentall, T. R. & Hogan, D. E. 1975 Concept learning in the pigeon: transfer to new matching and nonmatching stimuli. *Am. J. Psychol.* **88**, 233–244.

Zentall, T. R. & Hogan, D. E. 1978 Same/different concept learning in the pigeon: the effect of negative instances and prior adaptation to transfer stimuli. *J. exp. analyt. Behav.* **30**, 177–186.

Zentall, T. R., Hogan, D. E. & Edwards, C. A. 1984 Cognitive factors in conditional learning by pigeons. In *Animal cognition* (ed. H. L. Roitblat, T. G. Bever & H. S. Terrace), pp. 389–405. Hillsdale, N.J.: Lawrence Erlbaum Associates.

H. B. Barlow, F.R.S. (*University of Cambridge, U.K.*). My first question is, again, whether you think the learning tests you describe really test intelligence. If intelligence determines the knowledge and understanding an animal has of its normal environment (that is, the accuracy and completeness of its 'cognitive model'), would this have been tested in your experiments, or would your careful control of the conditions of your tests have more or less excluded the possibility of an animal using such background knowledge, acquired outside the learning situation?

My second question is methodological. For 40 years or so psychologists of sensation and perception have recognized that thresholds are statistical in nature and that when a human subject, or an animal, detects a weak signal he is often making remarkably efficient use of the information available. It is, furthermore, possible to express this efficiency quantitatively on an absolute scale (Rose 1944; Barlow 1980), and this makes possible valid comparisons between sensory performance at different tasks, between the performance of different species, and even between the performance of a single neuron and that of an intact whole animal.

Now learning is surely also a statistical task: some particular correlation, covariation, or association between two events has to be identified among all the possible such correlations, and this identification must be based on the statistical evidence the animal receives. It should be possible to apply the efficiency measure of statistical decision theory to this problem, and this would make comparisons across species and across tasks much more valid. Has this been done? I am aware that signal detection theory has been applied to measure the *sensory* capacities of animals, but has it been applied to their *learning* capacities?

To put the question another way, *rapid* learning is often equated with *better* learning, but this is not necessarily the case – the animal may be jumping to an incautious conclusion, just as a rash individual may achieve sensitivity by adopting a low criterion that leads to many false positive responses: have these considerations been taken into account by learning theorists?

References

Barlow, H. B. 1980 The absolute efficiency of perceptual decisions. *Phil. Trans. R. Soc. Lond.* B **290**, 71–82.
Rose, A. 1948 The sensitivity performance of the human eye on an absolute scale. *J. opt. Soc. Am.* **38**, 196–208.

N. J. Mackintosh. I suspect that there is no single process called 'intelligence'. Our argument was that one possibly important component of intelligence, and one that may differ between different animals, is the extent to which an animal can transfer an abstract rule across a change in physical stimuli. Our paradigm for studying this involves training animals on a task whose solution could be based on a general rule and testing them in a new situation to see how well they apply it. Although Professor Barlow is, I think, correct in suggesting that we try to control for or exclude the possibility of animals using knowledge acquired outside the learning situation, there is also a sense in which we are measuring precisely what he is asking us to. We are providing animals the opportunity to form a model of a particular environment, albeit an artificial one (that is, to learn that a particular rule describes what happens in it), and we are measuring the adequacy of that model to cope with a change in the environment. And our measures are certainly not equating rapid learning with better learning. Jackdaws, for example, learning matching discriminations very much more slowly than pigeons; and all corvids transferred to a new problem with a reversal of rule persisted in performing much less accurately than pigeons for hundreds of trials.

I do not know of attempts to apply signal detection theory specifically to learning capacities. The experimental paradigm which, as far as I can see, most clearly requires an animal to detect one imperfect correlation out of a variety of other correlations is 'probability learning'. An animal is trained on a discrimination between two alternatives, A and B, where choice of A is rewarded on, say 70 % of trials and choice B on the remaining 30 %. Even consistent choice of A will mean that only 70 % of the animal's first choices are rewarded, while purely random choice, or consistent choice of, say, the left-hand alternative, regardless of whether it is A or B, will ensure that 50 % of first choices are rewarded. The available evidence (Mackintosh 1974) suggests that pigeons are at least as efficient at this sort of task as rats, both animals eventually coming to select the most favourable alternative, A, on more than 90 % of trials (the question of whether they learn at the same speed has not been addressed). The task could obviously be made more difficult by arranging that one stimulus predicted reward on 70 % of trials, another on 65 %, a third on 60 % etc. It might be interesting to see whether this affected the outcome, but I know of no data that would answer this question.

Reference

Mackintosh, N. J. 1974 *The psychology of animal learning*. London: Academic Press.

Phil. Trans. R. Soc. Lond. B **308**, 67–78 (1985)
Printed in Great Britain

Actions and habits: the development of behavioural autonomy

By A. Dickinson

Department of Experimental Psychology, University of Cambridge, Downing Street, Cambridge CB2 3EB, U.K.

The study of animal behaviour has been dominated by two general models. According to the mechanistic stimulus–response model, a particular behaviour is either an innate or an acquired habit which is simply triggered by the appropriate stimulus. By contrast, the teleological model argues that, at least, some activities are purposive actions controlled by the current value of their goals through knowledge about the instrumental relations between the actions and their consequences. The type of control over any particular behaviour can be determined by a goal revaluation procedure. If the animal's performance changes appropriately following an alteration in the value of the goal or reward without further experience of the instrumental relationship, the behaviour should be regarded as a purposive action. On the other hand, the stimulus–response model is more appropriate for an activity whose performance is autonomous of the current value of the goal.

By using this assay, we have found that a simple food-rewarded activity is sensitive to reward devaluation in rats following limited but not extended training. The development of this behavioural autonomy with extended training appears to depend not upon the amount of training *per se*, but rather upon the fact that the animal no longer experiences the correlation between variations in performance and variations in the associated consequences during overtraining. In agreement with this idea, limited exposure to an instrumental relationship that arranges a low correlation between performance and reward rates also favours the development of behavioural autonomy. Thus, the same activity can be either an action or a habit depending upon the type of training it has received.

1. Teleology and intelligence

The study of the cause of behaviour has been dominated by two models of the animal. The mechanistic model assumes that at some level of causal analysis the occurrence of a particular activity is to be explained simply by appealing to the presence of an eliciting, releasing or triggering stimulus. Such a model underlies the physiologist's concept of a reflex, the ethologist's notion of a fixed action pattern and the psychologist's conditional and unconditional responses; indeed, so pervasive is the basic assumption of this model that it is common to refer to any behaviour as a 'response' and thus by implication, and often without any evidence, assume that there must be an eliciting stimulus.

This stimulus–response model of the animal is typically contrasted with a teleological account, according to which a particular behaviour occurs, not because it is elicited by a releasing or triggering stimulus, but rather because it is controlled at the time of performance by the animal's knowledge about the consequences of this activity. In other words, the teleological model claims that, at least, some behaviour is truly purposeful and goal-directed, and to distinguish such activities from 'responses', we refer to them as 'actions'. Of course, there is some truth in both these accounts: much behaviour is response-like in character, whereas there is little doubt that we, at least, are capable of goal-directed actions.

The relevance of this distinction between responses and actions for the general question of animal intelligence arises from the fact that a teleological system provides an animal with a much more flexible form of behavioural control. More specifically, a teleological system allows an animal to adjust its behaviour immediately and appropriately to changes in the value of its goals brought about either by an alteration in the animal's motivational state or by the acquisition of new knowledge about the value of the goals. I shall illustrate this important point by considering simple examples of both these types of goal revaluation.

In the case of motivationally mediated revaluation, let us assume that an animal has learned the routes to two water sources while thirsty, one of which has a much higher saline content than the other. The question of interest is whether, having had this experience, the animal could select the route to the saline goal when salt-deficient. Clearly the teleogical model potentially allows for such a selection; having knowledge about the consequences of taking each route, the animal could choose the one leading to saline when the value of this goal is enhanced. By contrast, a simple stimulus–response account would not permit appropriate selection. The training received while thirsty should simply have strengthened the capacity of the various stimuli along each route to elicit approach without providing the animal with any knowledge of the goals.

An analogous example can be constructed in the case of knowledge-based goal revaluation. In this case the animal might have learned two different routes to a particular food source. If subsequently the animal found out by following one route that the food source had become contaminated, a teleological, but not a stimulus–response process would allow the animal to avoid any further selection of the other route as well. It is clear from these examples that teleological processes greatly enhance the flexibility and power of an animal's cognition and the consequent control of behaviour, and that the possession of such processes gives a quantum jump in general intelligence above that exhibited by simple stimulus–response systems. It is, therefore, an important question in the field of animal intelligence to decide whether creatures other than ourselves are capable of true purposive actions. Answering this question is not the easy matter it might at first appear.

2. Goal revaluation

The teleological status of a particular activity can be determined rarely, if ever, by simple observation. The ethological and natural history literature has many examples of apparently purposeful and goal-directed behaviour that on simple experimental analysis turn out to be elicited responses. As a result, many years ago animal psychologists developed a variety of assays, that generally fall under the rubric of 'latent learning', for determining whether or not behaviour is controlled by knowledge about the goal. And I am sure that many psychologists would regard the question of whether mammals, at least as represented by the laboratory rat, are capable of true actions is an issue that has been resolved for a quarter of a century or more. As an anonymous referee of a recent paper on this problem observed, 'were he alive, Tolman would be perplexed that someone would still consider the issue in need of further experimentation'. But a closer inspection of the classic studies in this area reveals that the question is far from resolved.

This point can be illustrated by considering one of perhaps the most compelling demonstrations of apparent animal teleology in operation in an ingenious irrelevant-incentive study conducted by Krieckhaus & Wolf (1968). In effect, they implemented the assay for the teleological status of a behaviour suggested by the example of a motivationally induced change in the goal value that we have already considered. Two groups of thirsty rats were trained to press a lever, one for the sodium solution (Na) and the other for a potassium solution (K). Subsequently, the animals were sated for water and a sodium appetite induced before they were given an extinction test in which the rate of lever pressing in the absence of any reward was measured. Of course, if the animals trained with the sodium reward learn that lever pressing produces sodium, we should expect these animals to press more in the extinction test than those trained with the potassium reward.

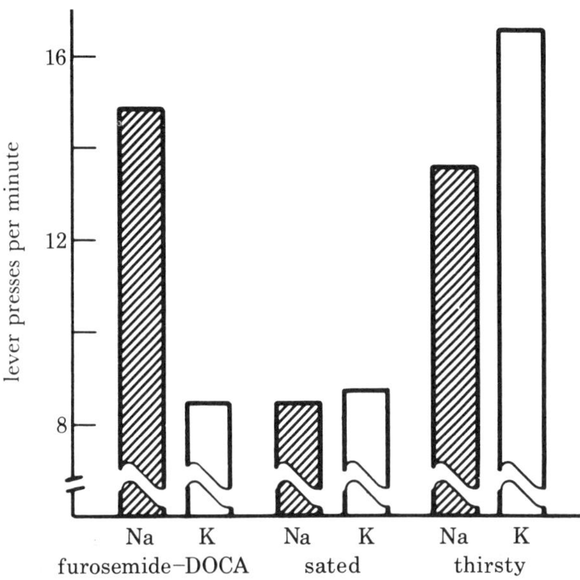

FIGURE 1. Mean lever press rates during the extinction test for groups trained with either the sodium (Na) or potassium (K) reward. Different groups were tested following the induction of a sodium appetite (furosemide-DOCA), sated for water and water-deprived.

That this prediction of the teleological account is fulfilled is shown in figure 1 which illustrates the results of a replication and extension of the Krieckhaus & Wolf (1968) study conducted in the Cambridge laboratory (Dickinson & Nicholas 1983). The rats trained with sodium pressed more in the extinction test than those rewarded with potassium when a sodium appetite was induced by the combined injection of the diuretic, furosemide, and desoxycorticosterone acetate (DOCA). This difference was not because the saline acted simply as a more effective reward. When other groups of rats were tested while they were either sated or thirsty, those trained with potassium performed at least as vigorously as those rewarded with sodium.

Plausible as the teleological account of this effect may seem, there is, in fact, a missing link in the empirical support. Krieckhaus & Wolf (1968) provided no evidence that the effect depends upon the animals having the opportunity to learn about the instrumental contingency or, in other words, the fact that lever-pressing causes saline delivery. If the same effect is seen whether or not the animals have the opportunity to learn about this contingency, the teleological account would be in trouble. To investigate the effect of the instrumental contingency, we trained two groups of rats to lever-press, one for the sodium solution (Na-W) and the other for, in this case, water (W-Na). In addition, however, both groups also received the other solution, but non-contingently or independently of lever-pressing. In fact, strictly speaking, these presentations were not non-contingent because the schedule was designed to minimize the possibility that the animals might believe that lever-pressing caused saline presentations owing to chance pairings (for details see Dickinson & Nicholas 1983). This means that both groups received the same number of saline presentations during training but only for one group (Na-W) did this solution act as a reward for lever pressing.

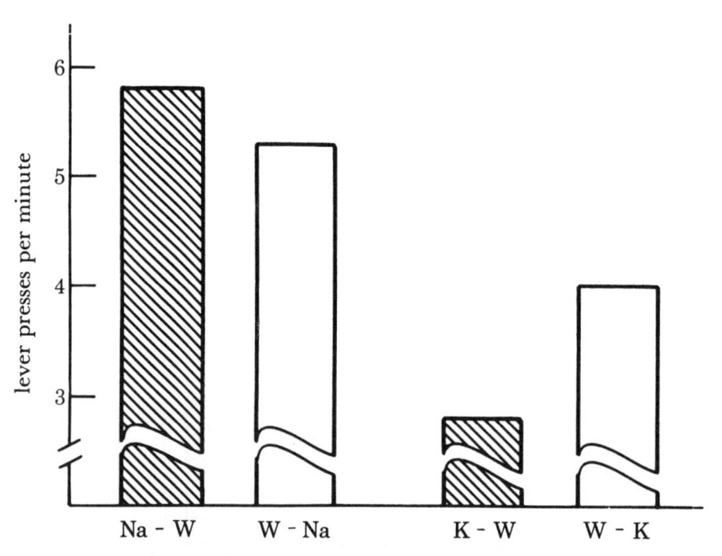

FIGURE 2. Mean lever press rates during the extinction test. Half of the groups were trained with either contingent sodium (Na-W) or potassium (K-W) and non-contingent water. The remaining groups received contingent water and either non-contingent sodium (W-Na) or potassium (W-K).

Given this training, the teleological account would anticipate essentially the same result as in the first study with animals that had the opportunity to learn that lever-pressing produced the saline (Na-W) performing more rapidly on test than those with no such opportunity (W-Na). In fact, we have consistently failed to detect such a difference when the amount of exposure to the sodium solution is equated as in the present study. Although group W-Na pressed at a slightly lower rate than group Na-W (see figure 2), this difference did not approach significance in the present study nor in other replications. Our failure to detect an effect of the instrumental contingency is not simply because the basic Krieckhaus and Wolf effect cannot be demonstrated with behaviour acquired on our complex schedule. Two further groups, groups K-W and W-K, were trained under exactly the same conditions as groups Na-W and W-Na, respectively, except that the potassium solution replaced the saline. When tested under a sodium appetite, groups K-W and W-K pressed significantly more slowly than the animals that received sodium during training. Any prior experience of saline, whether or not it is under the animal's control, elevates subsequent performance in the presence of a sodium appetite.

So what appeared to be a classic example of teleological control, on further experimental analysis turns out to be nothing of the sort. And, as far as I know, all the other previous demonstrations of latent learning either employ a behaviour, such as runway or maze performance, whose instrumental status is ambiguous (Mackintosh 1983), or fail to show that the effect depends upon the instrumental contingency. This is not the place to develop an explanation of the Krieckhaus and Wolf effect, although it is worth noting that our results would have been anticipated by a sophisticated development of the stimulus–response model, two-factor theory (Rescorla & Solomon 1967; Trapold & Overmier 1972).

Our failure to detect teleological control in response to a motivationally induced change in the value of the training goal may imply that the rat's behaviour is not under such control. Alternatively, lever-pressing may be an action, but one based upon associative knowledge that does not encode the aspect of the goal, in this case its saltiness, that is changed by motivational manipulation. I favour the latter interpretation for there is good evidence that this type of behaviour can be brought under teleological control using the second technique for changing the value of a goal, namely the acquisition of new knowledge.

Some years ago Christopher Adams and I (Adams & Dickinson 1981 *a*) trained hungry rats to lever press for one type of food, either sugar or mixed diet food pellets, while they received the other type non-contingently on the schedule employed in the previous study. We then devalued (D) the contingent or goal food in one group of rats, (D-N), while maintaining the value (N) of the non-contingent food. By contrast, the non-contingent food was devalued in the other group (N-D). This was done by giving the animals access to meals of the contingent and non-contingent food in the absence of the lever on alternate days after the lever press training had been completed. Animals in group D-N were averted from the contingent food by making them mildly ill immediately after they had consumed each meal of this food. The illness was induced by an injection of lithium chloride. Group N-D was similarly averted from the non-contingent food.

If lever pressing is an action, we should expect animals for which the contingent food is devalued to perform less vigorously when given the opportunity to lever-press again than those averted from the non-contingent food. In line with this prediction, figure 3 shows that group D-N pressed at a lower rate in the extinction test than group N-D. This difference was not because the aversion procedure had been more successful for the contingent than for the

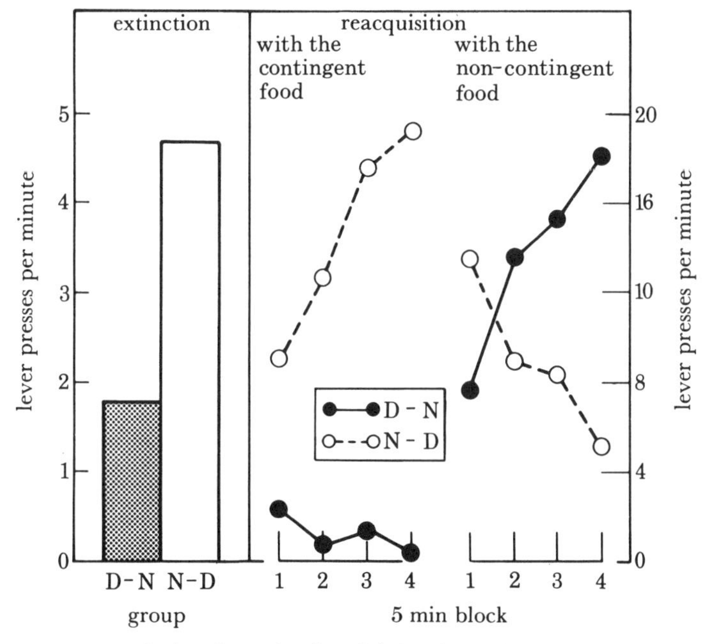

FIGURE 3. Mean lever-press rates during the extinction (left-hand panel) and reacquisition tests (right-hand panel) following the devaluation of either the contingent (group D-N) or non-contingent food (group N-D).

non-contingent food, which was in any case counterbalanced for type. Reacquisition tests demonstrated that both the contingent food in group D-N and the non-contingent food in group N-D had lost their capacity to act as rewards. The right-hand panel of figure 3 shows that during the first 20 min reacquisition test when lever-pressing produced the previously contingent food, performance was not re-established in group D-N, the group averted from this food. When a switch was made to the non-contingent food in the second reacquisition session, the decline in lever pressing by group N-D indicates that the non-contingent food had also lost its rewarding property for these animals.

I think that there can be little doubt, given this evidence, that the laboratory rat fits the teleological model; performance of this particular instrumental behaviour really does seem to be controlled by knowledge about the relation between the action and the goal. This conclusion will surprise few people. What might be more surprising is that the issue was not properly settled long ago.

3. Habits and the development of behavioural autonomy

The fact that an animal is capable of goal-directed actions does not imply that all its behaviour is under teleological control. Even the most purposive animals, such as ourselves, exhibit response-like behaviour. What then determines whether a particular behaviour is an action or a response? One might suppose that it is an immutable property of a particular behaviour to be either a response or an action, but a moment's reflection shows that this is probably not so. Perhaps the most obvious case in which apparently the same activity can be both an action and a response at different times is that of habit formation. The popular account of habit formation is that an instrumental behaviour, which starts out as an action controlled by knowledge about its relation to the goal, with repeated practice becomes a response, autonomous of the current value of the goal and simply triggered by the stimuli in whose presence it has been repeatedly performed. The response-like character of much of our own well-practised behaviour is revealed by what have been called 'slips of action' (for example, Norman 1981) that occur as a result of life's goal devaluation. Norman (1981), for example, quotes William James' (1890) well-known claim that 'very absent-minded persons in going to their bedroom to dress for dinner have been known to take off one garment after another and finally to get into bed, merely because that was the habitual issue of the first few movements when performed at a late hour'. For the middle class in the 19th century, the stimuli of the bedroom at that late hour tended to trigger a going-to-bed response even though its goal was inappropriate on that occasion or, in other words, devalued.

In spite of the strong anecdotal evidence that habit formation results from repeated practice, to my knowledge there is little or no experimental evidence in either humans or other animals that extended training renders behavioural control autonomous of the current value of the goal. Recently, however, Adams (1982) investigated the process of habit formation in the rat. By using our standard procedure, he trained two groups of rats to lever-press for sucrose pellets. One group was given a small amount of training, being allowed to perform only 100 rewarded lever-presses, whereas 500 presses were rewarded for the other, over-trained group. In both cases training was conducted at the rate of 50 rewards per session. We anticipated that pressing might have become a habit for the 500-press group through extended training and thus would be unaffected by goal devaluation. By contrast, this activity should have remained an action under the control of the current value of the goal for the 100-press group.

For half of the animals in the two training groups the reward was devalued (D) by pairing

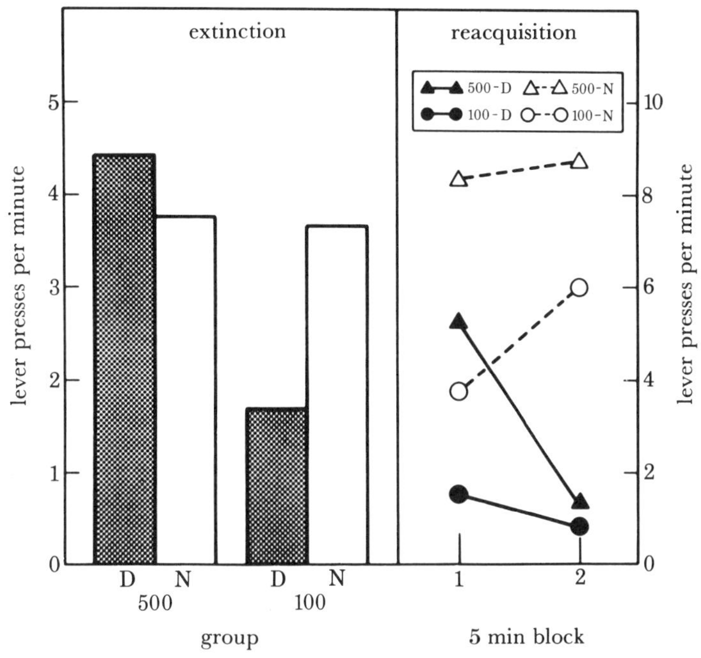

FIGURE 4. Mean lever-press rates during the extinction (left-hand panel) and reacquisition tests (right-hand panel) following either the devaluation (D) or non-devaluation (N) of the reward. The groups were allowed to make either 100 (groups 100-D and 100-N) or 500 rewarded lever presses (groups 500-D and 500-N) during training.

the consumption of sucrose pellets with the induction of illness, as in the previous experiment. The remaining rats acted as control animals for whom the reward was not devalued (N), although they received the same number of both meals of sucrose pellets and lithium chloride injections as the experimental animals, but on separate days. The left-hand panel of figure 4 shows that in a subsequent extinction test the group for whom the reward was devalued following 100 presses (100-D) performed less vigorously than the appropriate control group (100-N). Clearly, after a moderate amount of training, the lever-press remains an action under the control of the current value of the goal. By contrast, the outcome of the extinction test was completely different after over-training. After 500 rewards the rate of pressing by rats averted from the sucrose pellets (500-D), if anything, was greater than that of rats for whom the reward was not devalued (500-N).

This apparent difference in sensitivity to goal devaluation was not due to the ineffectiveness of the aversion procedure for the over-trained animals. The first 5 min block of a reacquisition test, not surprisingly, reproduced the pattern of result seen in the extinction test. Performance during the second block, however, shows that the sucrose pellets would no longer sustain lever pressing in group 500-D, indicating that this food had completely lost its rewarding property for these animals. Over-training can transform an action into a simple habit that is relatively autonomous of the current value of its original goal.

4. BEHAVIOUR–REWARD CORRELATION

The results of this study suggest that there might be some truth to the popular supposition that repetition produces habits. But it is not at all obvious that the failure of the animals allowed to make 500 presses during training to change their behaviour following goal devaluation was simply because they had pressed the lever so often. In the study comparing the effects of devaluing contingent and non-contingent rewards (page 71), the animals performed at least 500 presses and yet the behaviour remained sensitive to reward devaluation. This observation led me to consider the other factors that are changed by overtraining.

I approached this problem by attempting to track the animal's changing experience as training is extended. Figure 5 shows the acquisition function for the animals trained for 100

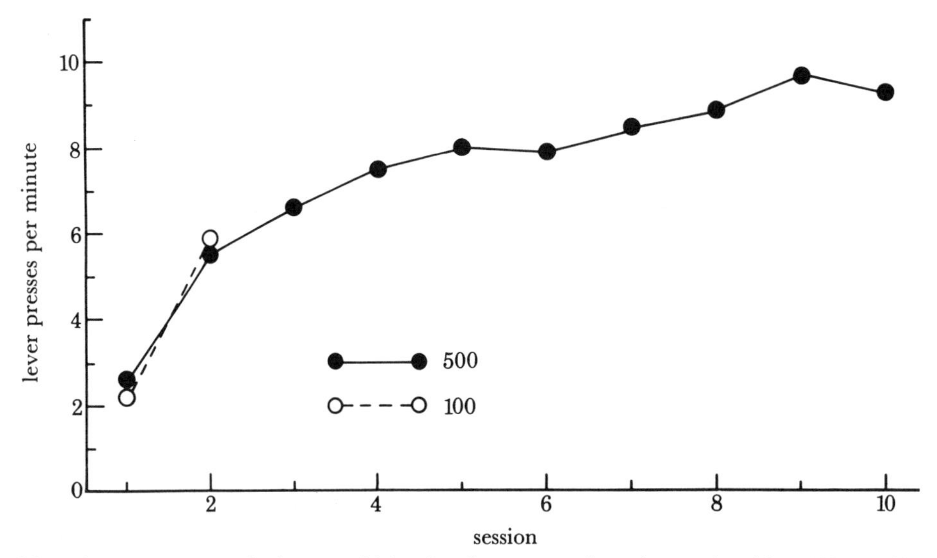

FIGURE 5. Mean lever press rates during acquisition for the groups allowed to make either 100 or 500 rewarded lever presses.

and 500 rewards. Extended training produced the typical negatively accelerated acquisition function in that the animals showed relatively large changes in the rate of performance across the initial sessions, but only small changes during the later sessions. This means that the animals experienced the relation or correlation between the rate of pressing and the reward rate over a large range of values during the first few sessions, but over only a restricted range later in training.

This point is illustrated more directly in figure 6 which portrays the relation between behaviour rate and reward rate. For the type of reward schedule used in the previous study in which each press produces a reward, a ratio schedule, there is, of course, a perfect correlation between behaviour and reward rates as represented by the linear function in figure 6; the faster that the animal performs the higher the rate at which the rewards occur. The points on this schedule function represent the relative average behaviour and reward rates on each session derived from the acquisition functions shown in figure 5. As can be seen, the animals in the over-training groups experienced the correlation between behaviour and reward rates over a wide range of values during the first five sessions, whereas their experience was limited to a much narrower range in the second five sessions. In fact, the variation in the animals' performance is so restricted during the later stages of training that they make little contact with the instrumental contingency which, of course, is defined by the way in which the occurrence of the reward varies with behaviour.

If we assume that an animal's performance is controlled by knowledge that directly reflects its current experience, then the development of behavioural autonomy follows directly from

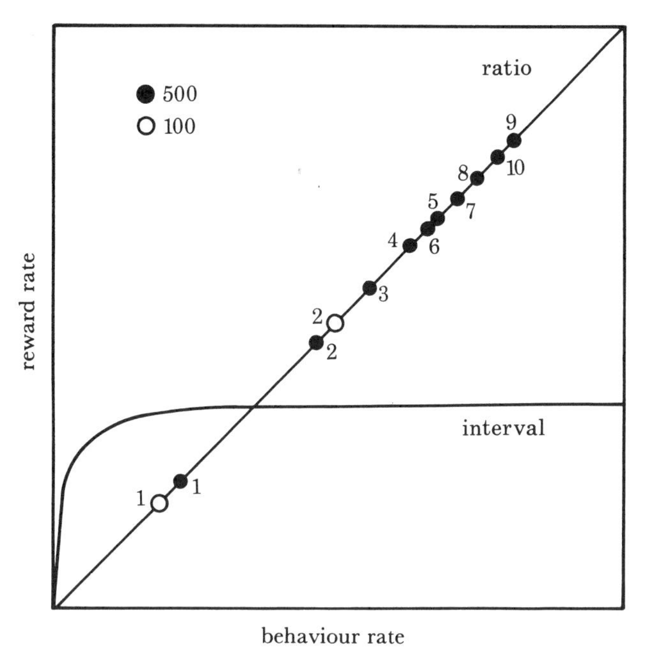

FIGURE 6. Reward rate as a function of the behaviour rate for ratio and interval schedules. The points on the ratio function represent the relative performance in each session of training for the groups allowed to make either 100 or 500 rewarded lever presses.

this analysis. When the animal's performance is varying during the early stages of training, it experiences, and thus stores knowledge about, a strong relation between behaviour and reward rates. Consequently, devaluing the reward depresses performance via this knowledge. By contrast, when the animals are over-trained, they no longer experience the behaviour–reward correlation with the result that their performance is no longer controlled by knowledge about this relation. In the absence of such knowledge, reward devaluation can have no effect and a habit has been established.

A clear prediction follows from this account of habit formation. If we found some other way of minimizing the animal's experience of a strong behaviour–reward correlation while maintaining performance, behavioural autonomy should result. A second type of reward schedule, an interval as opposed to a ratio schedule, provides such a procedure. Whereas on a random ratio schedule there is a fixed probability that each press will be rewarded, a random interval schedule arranges that a reward will become available with a constant probability in each time unit, for instance in each second. This reward then remains available until the next lever press collects it. Unlike a ratio schedule, which arranges a linear behaviour–reward rate function, this relation varies with the behaviour rate under an interval schedule. As figure 6 shows, at low levels of performance there is a strong positive relation between the behaviour and reward rates. As the behaviour rate rises, however, this relation weakens rapidly so that at modest levels of performance the reward rate is relatively unaffected by variations in the behaviour rate. This means that with minimal training on a random interval schedule the animal should no longer experience nor encode a strong behaviour–reward relation. Thus, an activity should be established as a habit much more rapidly on an interval schedule than on a ratio schedule.

To test this prediction, we (Dickinson *et al.* 1983) gave a group of rats limited training on random interval schedules. During the final training session these animals pressed on average at a rate of 12.4 per minute, a rate well within the range across which variation in performance has little effect on reward rate for this schedule. When the computer procedure that actually controlled the random interval schedule used during the final training session was driven at different rates by a mechanism that produced artificial lever-presses randomly in time, it was found that a fivefold increase in performance from 5 to 25 presses per minute only increased the reward rate, on average, from 0.92 to 1.06 pellets per minute. This 15% increase contrasts with a 400% rise in reward rate that would be produced by the same increment in performance on a ratio schedule. Consequently, we should expect lever pressing to be more readily established as a habit following this interval training than following an equivalent amount of training under a ratio contingency.

To check that any behavioural autonomy observed in these animals was owing to the lack of a behaviour–reward rate correlation rather than some other feature of the training, another group of rats were given the equivalent amount of training in terms of the number of rewards received on a ratio schedule. Furthermore, an attempt was made to match other features of the ratio and interval training, namely the probability that a press would be rewarded and the reward rate, by using a yoking procedure (for details see Dickinson *et al.* 1983). On the whole the matching was successful in that there was no significant difference between the reward probability in the two conditions, although the reward rate was, on average, 28% higher during the ratio training. I think, however, that we can be fairly confident in attributing any difference in behavioural autonomy following ratio and interval training to the behaviour–reward rate correlation arranged by the two schedules rather than some other feature of the contingencies.

The pattern of performance in an extinction test following reward devaluation by the food-aversion procedure confirmed our expectation. The left-hand panel of figure 7 shows that we replicated the basic reward devaluation effect following ratio training (R) in that animals (R-D) for whom the reward had been devalued (D) pressed at a lower rate than the control rats (R-N) for whom the reward was not devalued (N). More importantly, figure 7 also shows that performance following interval training (I) was impervious to reward devaluation; there was no detectable difference in the rate of pressing by animals for whom the reward had (I-D) and had not been devalued (I-N). The fact that the control animals trained on the ratio schedule (R-N) pressed more vigorously than those trained on the interval contingency (I-N) is a well-documented finding, although its explanation is currently a matter of dispute. The results of the reacquisition test again show that the extinction performance cannot be explained in terms of the differential effectiveness of the reward devaluation procedure. For both animals trained on the ratio (R-D) and interval schedules (I-D) the devalued sugar pellets would no longer act as an effective reward (see right-hand panel of figure 7).

These results suggest that, contrary to popular belief, habit formation is not a simple consequence of over-training or practice. Rather it appears to arise because over-training typically tends to reduce the variation in behaviour and thus the animal's experience of the relation that controls actions, namely the behaviour–goal correlation. Other ways of preventing the animal experiencing this relation once performance is established, such as training on an interval schedule, have the same effect. In the absence of the relevant experience, it is perhaps not so surprising that performance is no longer controlled by knowledge about the behaviour–goal correlation.

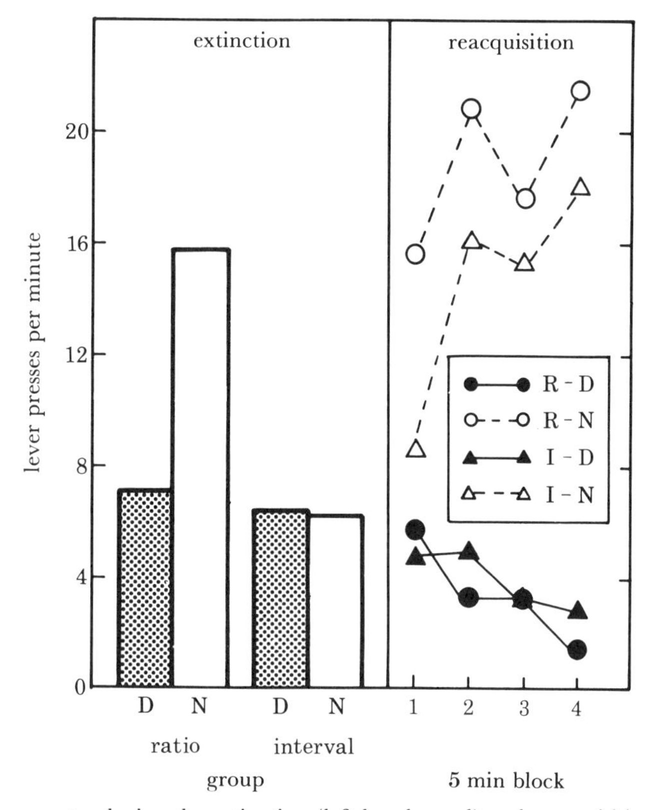

FIGURE 7. Mean lever press rates during the extinction (left-hand panel) and reacquisition tests (right-hand panel) following either the devaluation (D) or non-devaluation (N) of the reward. The groups were trained on either a ratio (groups R-D and R-N) or an interval schedule (groups I-D and I-N).

5. CONCLUSIONS

The conclusion drawn from the classic studies of latent learning was correct even if the outcome of these experiments did not necessarily justify this claim. The selective effect of devaluing the contingent reward on performance demonstrates that the laboratory rat, at least, is capable of purposive goal-directed actions. Moreover, our findings show that instrumental behaviour does not represent a homogeneous category. By varying the training conditions appropriately, the same behaviour in the same species can both show behavioural autonomy and be brought under teleological control. This conclusion has implications for the study of comparative intelligence. For instance, we do not yet know whether the two types of control can exist independently of each other or whether they are both based upon a common learning process. If the former possibility is true, comparative psychology is not faced just with determining whether the species of interest is capable of instrumental conditioning, but also with the problem of identifying the type of instrumental control.

From my point of view, however, the main implication of our results concerns the nature of the cognitive processes controlling instrumental behaviour. My colleagues and I (Adams & Dickinson 1981 *b*; Dickinson 1980; Mackintosh & Dickinson 1979) have argued that teleological control of instrumental behaviour cannot be explained, at least at the psychological level, in terms of internal associations which have just excitatory or inhibitory properties. Rather, we argue that the knowledge about the action–goal relation must be encoded in a propositional-like form so that it can be operated on by a practical inference process to generate the instrumental performance. In this sense actions are inherently rational in a way that responses can never be.

The work reported in this paper was supported by a grant from the S.E.R.C. I would like to thank N. J. Mackintosh, D. Charnock and C. Preston for their comments on an earlier draft of this manuscript.

References

Adams, C. D. 1982 Variations in the sensitivity of instrumental responding to reinforcer devaluation. *Q. Jl exp. Psychol.* **34** B, 77–98.

Adams, C. D. & Dickinson, A. 1981*a* Instrumental responding following reinforcer devaluation. *Q. Jl exp. Psychol.* **33** B, 109–112.

Adams, C. D. & Dickinson, A. 1981*b* Actions and habits: Variations in associative representations during instrumental learning. In *Information processing in animals: memory mechanisms* (ed. N. E. Spear & R. R. Miller), pp. 143–165. Hillsdale, New Jersey: Erlbaum.

Dickinson, A. 1980 *Contemporary animal learning theory*. Cambridge: Cambridge University Press.

Dickinson, A. & Nicholas, D. J. 1983 Irrelevant incentive learning during instrumental conditioning: the role of the drive-reinforcer and response-reinforcer relationships. *Q. Jl exp. Psychol.* **35** B, 249–263.

Dickinson, A., Nicholas, D. J. & Adams, C. D. 1983 The effect of the instrumental contingency on susceptibility to reinforcer devaluation. *Q. Jl exp. Psychol.* **35** B, 35–51.

James, W. 1890 *The principles of psychology*. New York: Holt.

Krieckhaus, E. E. & Wolf, G. 1968 Acquisition of sodium by rats: Interaction of innate mechanisms and latent learning. *J. comp. physiol. Psychol.* **65**, 197–201.

Mackintosh, N. J. 1983 *Conditioning and associative learning*. Oxford: Oxford University Press.

Mackintosh, N. J. & Dickinson, A. 1979 Instrumental (type II) conditioning. In *Mechanisms of learning and motivation* (ed. A. Dickinson & R. A. Boakes), pp. 143–169. Hillsdale, New Jersey: Erlbaum.

Norman, D. A. 1981 Categorization of action slips. *Psychol. Rev.* **88**, 1–15.

Rescorla, R. A. & Solomon, R. L. 1967 Two-process learning theory: relationship between Pavlovian conditioning and instrumental learning. *Psychol. Rev.* **74**, 151–182.

Trapold, M. A. & Overmier, J. B. 1972 The second learning process in instrumental learning. In *Classical conditioning*. Vol. II. *Current research and theory* (ed. A. H. Black & W. F. Prokasy), pp. 427–452. New York: Appleton-Century-Crofts.

Phil. Trans. R. Soc. Lond. B **308**, 79–86 (1985) [79]
Printed in Great Britain

The temporal context of spatial memory

By D. S. Olton

Department of Psychology, The Johns Hopkins University, Baltimore, Maryland 21218, *U.S.A.*

Foraging strategies provide an effective and convenient means of investigating some characteristics of animal intelligence. The temporal and spatial distributions of food in the environment are not entirely random, and animals that use information about these patterns can gain an adaptive advantage over animals that do not. For some types of strategies to be effective, the animal must remember the temporal context of a visit to a spatial location and use this memory to make decisions about the current distribution of food in the environment. Experiments with different species (rat, Hawaiian honeycreeper, Marsh tit, Clark's nutcracker) have used variations of a delayed conditional discrimination to examine the cognitive processes influencing this type of memory. These include: primacy, recency, proactive interference, retroactive interference, decay, consolidation, and chunking. The results of these experiments provide information about the types of memory processes that animals use when searching for food, and illustrate the usefulness of combining psychological and ethological approaches when studying animal intelligence.

Every event that we wish to remember happens in a particular place at a particular time. For some types of decisions, all three of these components (the event, its spatial location, and its temporal context) must be associated together if their memory is to help the individual make the correct choice.

Animals are frequently confronted with these types of decisions in their natural environments (Krebs 1978). Resources are not distributed homogeneously. In any particular place, more resources are available at some times than at others. At any particular time, more resources are available in some places than at others. Furthermore, these temporal and spatial patterns are cyclical and predictable. Thus, a memory of the spatial and temporal context of previously experienced events can be useful in making decisions about the current distribution of resources.

Being in the right place at the right time confers an adaptive advantage. The animal increases the availability of resources, and decreases the cost of obtaining them. Thus, environmental pressures encourage the development of a memory system that can associate an event with its spatial and temporal context.

Environmental pressures favouring the development of an ability do not ensure that animals will indeed acquire this ability. The fossil record stands in mute but elegant testimony to the great probability of failure. However, empirical data demonstrate that many animals have developed this type of memory system. Convergent evolution in different species is strong evidence of a consistent environmental pressure, and this memory helps animals behave adaptively (that is, intelligently) in many different situations. My first example is from a laboratory discrimination task with the ubiquitous *Ratus norwegicus*. The subsequent examples are from naturalistic settings (real or simulated) with several other species. In all of these, the analysis involves answering the same three questions:

(i) is a memory of the temporal and spatial context of an event useful for accurate decisions about the current distribution of resources?

(ii) Do animals use this memory?

(iii) If animals do use this memory, what are the characteristics of it?

1. Radial arm maze

The test apparatus is a maze with arms radiating from a central platform. At the start of each trial, one pellet of food is placed at the distal end of each arm. Food is not replaced during a trial. Consequently, the optimal strategy for the rat is to visit each arm once and only once during a trial (Olton & Samuelson 1976).

A memory of the temporal context of a visit to each arm can help the rat to determine which arms currently have food. If an arm has not been visited during the current trial, that arm has food and the rat ought to go to the end of it. If an arm has been visited during the current trial, that arm no longer has food and the rat ought not to go to the end of it. Only by determining whether an arm has been visited during the *current* trial can a rat decide whether or not it has food. Visits to arms in previous trials must be distinguished from visits to arms in the current trial. A memory of the temporal context of each visit can help make this discrimination.

However, many relevant redundant strategies can be used to determine where food is located. These strategies use different discriminative stimuli and different types of memory. They are relevant because they can be used to choose correctly. They are redundant because with the usual test procedure the experimenter is unable to determine the extent to which each one actually directs choice behaviour. For example, intramaze stimuli from the food or from previous visits, and specific response patterns, can all lead to correct choices.

When relevant redundant strategies are available in a discrimination procedure, two steps can be taken to determine if an animal actually uses the particular strategy that the experiment wishes to study. First, the strategy of interest is made independent of all others. Second, only that strategy is made relevant. A comparison of choice accuracy in the modified task with that in the original task indicates the extent to which the animals can use only the single strategy to choose correctly.

Experiments with the radial arm maze were designed to force the rats to remember the temporal context of a visit to each food location and prevent them from using other strategies to determine where food was located. The rats still chose among the arms very accurately. Consequently, rats do use the memory of the temporal context of their visit to each spatial location to decide where food is located (Beatty & Shavalia 1980*a*; Maki *et al.* 1985; Olton *et al.* 1977; Olton & Samuelson 1976; Roberts 1979, 1981; Roberts & Dale 1981; Roberts & Smythe 1979; Walker & Olton 1979). Similar conclusions have been obtained from many other experiments with mazes (O'Keefe & Conway 1978; Olton 1979).

Identifying the characteristics of this memory requires close attention to experimental procedures and the logic of interpretation. Although the choices themselves can be observed and quantified without difficulty, using them to infer the characteristics of memory must proceed cautiously. For example, consider the problems that arise if response patterns are not eliminated. When leaving an arm and returning to the central platform, a rat usually turns in a particular direction (clockwise or counter-clockwise) and goes to an arm close to the one previously visited. Although these response patterns do not constitute a strategy that is required to choose correctly, they are still habits and bias choice behaviour (Olton *et al.* 1977; Roberts & Dale 1981). Inferences about memory are compromised because the experimenter cannot determine the extent to which any given visit is influenced by memory or by the variables controlling the response pattern. In short, the extent to which choice behaviour can be used to draw conclusions about memory is limited by the extent to which memory and only memory influences that choice behaviour.

Consequently, the following discussion of the characteristics of memory is based on only those experiments that used procedures designed to maximize the influence of memory on choice behaviour. In most cases, the relative salience of extramaze cues was enhanced by having an open maze in a well-lighted room with many distinctive extramaze stimuli. Sides were often placed along the edges of each arm to help the rat stay on the maze, but these were low enough to produce only a minimal impediment of the rat's ability to see the extramaze cues. In the best experiments, response patterns were prevented by a series of forced visits followed by a choice between two arms. For each forced visit, the rat was given access to only a single arm. Consequently, the experimenter, rather than the rat, chose which arms were to be visited and the order of those visits. After these visits, the rat was given access to two arms and allowed to choose between them. In other experiments, the rat was confined to the centre platform after each visit, or removed from the maze during the trial.

Primacy

Primacy is more accurate memory for items at the beginning of a list than for items in the middle of the list. With the usual win–shift response–reinforcement contingency described above, primacy did not occur. The probability of making an error (returning to an arm visited previously during a trial) was no less for the first arm visited that it was for arms visited in the middle of the trial (DiMattia & Kesner 1985; Kesner & Novak 1982; Olton & Samuelson 1976; Roberts & Smythe 1979). With a win–stay response–reinforcement contingency, however, primacy did occur; the probability of making an error to the first arm visited during a trial was less than that to arms that were visited in the middle of that trial (DiMattia & Kesner 1985).

Recency

Recency is more accurate memory for items at the end of a list than for items in the middle of the list. Recency did occur. The probability of making an error to an arm just recently visited was less than that for arms visited in the middle of a trial (DiMattia & Kesner 1985; Kesner & Novak 1982; Olton & Samuelson 1976; Roberts & Smythe 1979).

Proactive interference

Proactive interference is the deleterious effect of previously learned information on the ability to remember currently learned information. Proactive interference occurred within a trial on the same maze, between completed trials on the same maze, and between uncompleted trials on different mazes when all testing was completed in a relatively short period of time (less than 15 min). The probability of making an error increased as the number of arms visited before the one in question increased. The magnitude of the effect was small (if present) after a few visits. It was positively accelerated so that the magnitude of the effect was substantial after many visits (Roberts 1981; Roberts & Dale 1981). Interference was not produced by stimuli unrelated to those around the maze (Maki *et al.* 1979).

Retroactive interference

Retroactive interference is the deleterious effect of currently learned information on the memory of previously learned information. Retroactive interference occurred within a trial on the same maze, and between uncompleted trials on different mazes. The probability of making an error increased as the number of visits following the initial visit to an arm increased. The magnitude of the effect was minimal (if present) after a few visits and was positively accelerated (Beatty & Shavalia 1980*b*; Maki *et al.* 1979; Roberts 1981). Interference was not produced by stimuli unrelated to those around the maze (Maki *et al.* 1979).

Decay

Decay is the progressive deterioration of the memory with the increasing passage of time following the event to be remembered. Decay can never be completely isolated from interference due to events because events always occur as time passes. In these experiments, however, each rat was placed in a holding cage for the delay interval. Thus, the intervening events were minimized and were not visits to arms on mazes as in the previously described studies examining interference. Decay occurred. The probability of making an error increased as the time from the initial visit to that arm increased. This effect was minimal (if present) with delays of up to several hours and was positively accelerated (Beatty & Shavalia 1980*a*, *b*; Gaffan & Davies 1981; Knowlton *et al.* 1985; Roberts 1981; Roberts & Dale 1981; Roberts & Smythe 1979).

Consolidation

Consolidation is the decreased susceptibility of memory to disruption as the length of time following the information to be remembered increases. Consolidation did not occur within 8 h. The probability of making an error was just as great when 8 h intervened between the initial visit to an arm and the disruptive event (electroconvulsive shock or intrahippocampal stimulation that produced seizures) as when only a few seconds intervened (Knowlton *et al.* 1985; Shavalia *et al.* 1981).

'Chunking'

'Chunking' is the organization of information to be remembered so that it can be remembered more efficiently. A retrospective memory of the arms that have been chosen may be used at the beginning of a trial, while a prospective memory of the arms that are to be chosen may be used at the end of a trial. This shift from a retrospective memory to a prospective memory can reduce the number of visits that have to be remembered at any one time (although it adds the additional requirement of determining which type of memory is currently being used) (Cook *et al.* 1985). Organization of visits on the basis of the spatial location of the arms did not occur (Olton & Samuelson 1976).

2. Naturalistic settings

In each of the following examples, an animal was tested with commonly consumed food in either the natural habitat or a simulation of it. The discrimination problem facing the animal was one in which the memory of the temporal context of a visit to a spatial location could help determine the current location of food. The animals did use this type of memory to help decide which location should be visited: cues from the food itself or response patterns were not necessary for accurate choices. The controls to emphasize the use of memory and describe its characteristics were not as stringent as in the experiments with rats, and only some of the characteristics of the memory used by these animals have been identified.

Loxops virens

The Hawaiian honeycreeper, *Loxops virens*, obtains nectar from flowers of the Mamane tree, *Sophora chrysophylla*. During each visit to a flower, the bird removes all the nectar. The flower takes several hours to replace this nectar. During this time, the optimal strategy for the bird is to avoid just previously visited flowers and search elsewhere. Although the accuracy of choice behaviour of the birds varied from day to day, the probability of returning to a previously visited cluster of flowers was slightly less than that expected by chance. Substantial forgetting took place for visits to flowers several hours after the visit occurred. The variables responsible for this forgetting were not identified (Kamil 1978).

Parus palustris

Marsh tits, *Parus palustris*, store seeds in scattered locations and retrieve them many hours later. (This type of hoarding should minimize the probability of other birds stealing the seeds). After the birds have stored the seeds, the optimal strategy is to go to each location containing a seed once and only once, obtaining the seed from it on the first visit. A naturalistic environment was created in an aviary. Twelve small 'trees' were made from the limbs of real trees. Holes were drilled into the limbs. A piece of black cloth was stapled above each hole and could be lowered to cover the hole preventing the bird from seeing if a seed was located in the hole (Shettleworth & Krebs 1982).

At the beginning of each trial, a container of seeds was placed in the aviary. For the *hoarding* phase of the trial, the bird was released into the aviary and permitted to store these seeds. The bird was then removed from the aviary. For the *recovery* phase of each trial, the container of seeds was removed and the bird was allowed to search for seeds in the holes in the trees.

The choice accuracy of the birds during the recovery phase was very high, and remained significantly above chance even after 30 seeds had been obtained. This accuracy did not result from fixed search paths or discriminative stimuli from the food itself. The probability of a correct choice descreased as the number of choices during the recovery phase increased; because the interval between the hoarding phase and the recovery phase was long (several hours) compared with the interval for the storage of all the seeds in the hoarding phase (about 10 min), interference was more likely than decay.

Nucifraga columbiana

Clark's nutcrackers, *Nucifraga columbiana*, store many thousands of seeds from pine cones during the short time that these are available and then retrieve them during the subsequent year. After the hoarding of the seeds is completed, the optimal strategy for the nutcrackers is to visit each location of hoarded seeds, obtain the seeds there, and then not return. Observations of the birds in the natural environments suggest that they are very accurate in the choices of places to go to recover seeds (see review in Kamil & Balda 1985).

The aviary had a floor with 180 holes in it. In each hole was a cup full of sand. A wooden plug could be placed over the cup to prevent the bird from getting access to the cup. Rocks and boards were distributed around the floor to provide stimuli distinguishing the spatial locations of the different cups. In each of three *caching* sessions, a container of seeds was placed in the aviary and 18 of the holes were open. The bird was allowed to store seeds in only these sites. During the subsequent *recovery* session, the container was removed, all holes were opened, and the bird was allowed to choose among them (Kamil & Balda 1985).

The probability of going to a hole in which seeds had been placed remained above the levels expected by chance throughout the recovery session, but decreased more rapidly as the number of holes visited increased. Because the interval between the last caching session and the recovery session (ten days) was only slightly longer than the interval between the first and the last caching session (four days), this pattern may have been due to either interference or decay. When the birds made an error, they had a high probability of going to a hole adjacent to one in which seeds were stored.

3. Discussion

Animals do remember the temporal context of a visit to a spatial location and use this information to make decisions about the current distribution of resources in their environment. For rats solving a discrimination task in the laboratory, the parameters influencing the accuracy of this memory are essentially the same as those influencing human memory. For other animals in naturalistic settings, fewer parameters have been examined, but these also affect memory in the same way as they do in people.

These similarities suggest that analogous memory systems are present in all of these species. If such is the case, differences in the ability to remember information must result from quantitative variations in the shapes of the functions relating the parameters to memory rather than from qualitative differences in the types of parameters that influence memory.

These experiments indicate the advantage of combining two different perspectives: that commonly taken by psychologists when investigating memory in laboratory discrimination tasks, and that taken by ethologists when investigating optimal foraging in naturalistic settings. The psychological perspective provides information about the characteristics of memory and the types of experimental designs that can be used to investigate it. The ethological perspective provides information about the types of problems that animals face in their natural habitats and the ways in which memory can help solve them.

Historically, laboratory procedures have had more controlled experimental designs than those in naturalistic studies and these designs have permitted more definite conclusions about the variables that influence behaviour. However, laboratory experiments have often ignored the extent to which the task is meaningful to the animal being tested. Ultimately, every comparative study must justify its choice of task, and the major step in assessing the validity of any task is a comparison of its demands with those experienced by the animal in its natural habitat. Consequently, the most productive experiments are likely to be those that incorporate the advantages of experimental control into naturalistic settings. Some characteristics of the experiments reviewed in the last part of this chapter represent an excellent first step in this direction.

However, many questions remain to be answered. The representation of both the temporal (Church 1984; Gibbon *et al.* 1984) and the spatial codes (O'Keefe & Nadel 1978) remains to be identified, along with the ways in which these are associated together. A comparative study that goes beyond the experimental parameters influencing choice accuracy is necessary to determine the cognitive processes supporting this memory. Although the basic neuronal elements in the brains of birds and mammals are the same, their organization into ganglia and nuclei is very different. The question is whether the memory processes described for each of these species arise from the common neuronal elements or the different neuronal organizations. If the former, the psychological as well as the neurological mechanism involved in memory may be homologous, sharing the same origin. If the latter, both mechanisms are analogous, achieving the same goal through different means.

Finally, the spatial distribution of many resources other than food change with predictable temporal patterns, and those patterns can be much more complex than ones described here. Although relatively complicated schedules of reinforcement have been investigated in laboratory operant boxes, they have only begun to be studied in ethologically relevant settings. Thus, experiments need to address the broader issue of how animals allocate their time and energy among all the activities available to them. In this allocation, remembering the temporal context of a visit to a spatial location will help the animals choose intelligently among the various alternatives.

References

Beatty, W. W. & Shavalia, D. A. 1980*a* Spatial memory in rats: time course of working memory and effect of anesthetics. *Behav. neural Biol.* **28**, 454–462.

Beatty, W. W. & Shavalia, D. A. 1980*b* Rat spatial memory; resistance to retroactive interference at long retention intervals. *Anim. Learn. Behav.* **8**, 550–552.

Church, R. M. 1984 Properties of the internal clock. *Ann. N.Y. Acad. Sci.* **423**, 566–582.

Cook, R. G., Brown, M. F. & Riley, D. A. 1985 A flexible memory processing by rats: use of prospective and retrospective information in the radial maze. (Submitted).

DiMattia, B. V. & Kesner, R. P. 1985 Serial position curves in rats: Automatic vs. controlled information processing. (Submitted.)

Gaffan, E. A. & Davies, J. 1981 The role of exploration in win-shift and win-stay performance on a radial maze. *Learn. Motiv.* **12**, 282–299.

Gibbon, J., Church, R. M. & Meck, W. H. 1984 Scalar timing in memory. *Ann. N.Y. Acad. Sci.* **423**, 52–77.

Kamil, A. C. 1978 Systematic foraging by a nectar-feeding birds, the amakihi (*Loxops virens*). *J. comp. Physiol. Psych.* **92**, 388–396.

Kamil, A. C. & Balda, R. C. 1984 Cache recovery and spatial memory in Clark's nutcrackers (*Nucifraga columbiana*). *J. exp. Psychol.: Anim. Behav. Process.* (Submitted.)

Kesner, R. P. & Novak, J. M. 1982 Serial position curve in rats: Role of the dorsal hippocampus. *Science, Wash.* **218**, 173–174.

Knowlton, B., McGowan, M., Olton, D. S. & Gamzu, E. 1985 Hippocampal stimulation disrupts spatial working memory even after eight hours for consolidation. *Behav. neural Biol.* (Submitted.)

Krebs, J. R. 1978 Optimal foraging: Decision rules for predators. In *Behavioural ecology: An evolutionary approach* (ed. J. R. Krebs & N. B. Davies), pp. 23–63. Oxford: Blackwell Scientific Publications.

Maki, W. S., Beatty, W. W., Hoffman, N., Bierley, R. A. & Clouse, B. A. 1985 Evaluation of nonmemorial influences on rats' performance in the radial-arm maze with long retention intervals. *Behav. neural Biol.* (Submitted.)

Maki, W. S., Brokofsky, S. & Berg, B. 1979 Spatial memory in rats: Resistance to retroactive interference. *Anim. Learn. Behav.* **7**, 25–30.

O'Keefe, J. & Conway, D. H. 1978 Hippocampal place units in the freely moving rat: Why they fire where they fire. *Exp. Brain Res.* **31**, 573–590.

O'Keefe, J. & Nadel, L. 1978 *The hippocampus as a cognitive map.* Oxford: Oxford University Press.

Olton, D. S., Collison, C. & Werz, M. A. 1977 Spatial memory and radial arm maze performance of rats. *Learn. Motiv.* **8**, 289–314.

Olton, D. S. & Samuelson, R. J. 1976 Remembrance of places passed: Spatial memory in rats. *J. exp. Psychol.: Anim. Behav. Process.* **2**, 97–116.

Olton, D. S. 1979 Mazes, maps, and memory. *Am. Psych.* **34**, 583–596.

Roberts, W. A. 1979 Spatial memory in the rat on a hierarchical maze. *Learn. Motiv.* **10**, 117–140.

Roberts, W. A. 1981 Retroactive inhibition in rat spatial memory. *Anim. Learn. Behav.* **9**, 566–574.

Roberts, W. A. & Dale, R. H. 1981 Rembrance of places lasts: Proactive inhibition and patterns of choice in rat spatial memory. *Learn. Motiv.* **12**, 261–281.
Roberts, W. A. & Smythe, W. E. 1979 Memory for lists of spatial events in the rat. *Learn. Motiv.* **10**, 313–336.
Shettleworth, S. J. & Krebs, J. R. 1982 How marsh tits find their hoards: The roles of site preference and spatial memory. *J. exp. Psychol.: Anim. Behav. Process.* **8**, 354–375.
Walker, J. A. & Olton, D. S. 1979 The role of response and reward in spatial memory. *Learn Motiv.* **10**, 73–84.

Phil. Trans. R. Soc. Lond. B **308**, 87–99 (1985)   [87]
Printed in Great Britain

Hippocampus: memory, habit and voluntary movement

By D. Gaffan
Department of Experimental Psychology, South Parks Road, Oxford OX1 3UD, U.K.

A general method for studying monkeys' memories is to teach the animals memory-dependent performance rules: for example, to choose, out of two visual stimuli, the one that flashed last time the animal saw it. One may thus assess the animal's memory for any arbitrarily chosen event such as flashing even if the event itself has no intrinsic importance for the animal. The method also allows assessment of an animal's memory of the animal's own previous behaviour. The use of these methods has revealed a simple generalization about the function of the hippocampus in memory: hippocampal lesions impair memory of the voluntary movement that a stimulus previously elicited, but leave intact memory for relations between environmental events other than voluntary movements. The impairment in memory for voluntary movements produces deficits in exploration and in habit formation.

1. Memory tasks

Monkeys are adept at learning memory-dependent performance rules. A convenient method for teaching a monkey such a rule is to use a large population of visually discriminable stimulus objects. With the first pair of objects drawn from the population one proceeds to give one member of the pair one history, H1, and the other a different history, H2, in one or more 'acquisition trials' that are witnessed by the monkey. Subsequently the pair is presented for the animal's choice at a 'retention test'. Then that pair is discarded and the sequence of acquisition and retention is repeated with a new pair of objects. At retention tests it is always the object that has one of the histories, say H2, that is rewarded, and the other is not. Over a series of such tests the monkey will learn to choose between objects at retention tests according to their history. The accuracy of the choices made will therefore come to reflect the animal's ability to recall differentially H1 and H2, that is, to discriminate between the memories of H1 and H2. The stimulus populations used in the experiments reviewed below are either three-dimensional objects presented to the monkey by hand, or flat coloured patterns generated by a computer. The reward at retention tests is usually a peanut. The histories in question may be quite various, as table 1 exemplifies.

Some of the tasks in table 1 are well known and are given their conventional names in the table; for others a suggested name is given in quotation marks. The detailed methods of experiments with these tasks are described in the papers cited below in §2 and 3. For illustration in the table only the first four pairs of stimulus objects are shown for each task although in every case more retention tests than four would be required to teach a monkey the task. The two stimuli of each pair are identified by capital letters. The row of the table labelled 'retention test' shows the common retention test for all of the tasks, 'vs' indicating the permission to choose, '+' indicating the rewarded correct choice, and the order of the stimulus identifications indicating the left–right spatial position of the stimuli at the retention test. In the other rows of the table the acquisition trials for the different tasks are shown, separated by commas when there are more than one of them to be presented successively.

Table 1. Some memory-dependent performance rules

task name	pair number 1	2	3	4
matching to baited samples	A^+	C^+	F^+	H^+
matching to unbaited samples	A	C	F	H
non-matching to randomly baited	B^+	D	E^+	G
'sensory recall (f)'	Af, Bx	Dx, Cf	Ex, Ff	Hf, Gx
'congruent recall'	A^+, B	D, C^+	E, F^+	H^+, G
object discrimination learning set	A^+ vs B	D vs C^+	E vs F^+	H^+ vs G
'incongruent recall'	A, B^+	D^+, C	E^+, F	H, G^+
retention test (all tasks)	A^+ vs B	C^+ vs D	E vs F^+	G vs H^+

See the text for explanation of the symbols.

The first three tasks, matching and non-matching, require the monkey to answer a simple question: which object has been recently presented and which has not? The object that is presented in acquisition is known as the sample. In cases where the sample at acquisition is baited with a reward, that is indicated by '+'.

Sensory recall requires an answer to a more complicated question: which stimulus was associated with which of two possible target events to be recalled? The target events in this case are 'f' which means that the stimulus flashes on and off at acquisition, and 'x' which means that the stimulus expands at acquisition. I have shown the version (f) where it is the memory of the stimulus having flashed that signals reward at retention, but naturally the opposite rule can also be learned. Though one or other of the memories of 'f' or 'x' signals reward, 'f' and 'x' themselves (not the memories of them) are both equally predictive of non-reward, since neither ever occurs at retention. This is an important difference between sensory recall and the bottom three tasks of the table, where acquisition events are themselves either rewarding or non-rewarding.

Congruent recall is so-called because the memory of an event predicts, in this case, that event, the events in question being reward and non-reward. In incongruent recall, on the other hand, the opposite relation is seen, since in that case the memory of reward predicts non-reward and the memory of non-reward predicts reward. In object discrimination learning set (o.d.l.s.) there is the same congruency between memory and expectation as exists in congruent recall. Furthermore, in o.d.l.s. the acquisition trial is operationally identical to the retention test.

O.d.l.s. is a familiar task that has been studied over many years in a number of laboratories. It is apparently simple. In a typical o.d.l.s. experiment all that happens is that a series of single simple two-choice visual discrimination problems is presented: first A and B are presented for a number of trials with A consistently rewarded, then C and D are similarly presented with C rewarded, and so on; and, not surprisingly, the rate of within-problem learning improves with practice. This simplicity is deceptive. The reason why all within-problem trials look the same in o.d.l.s. is, first, because each trial performs two functions that need to be separated logically, as an acquisition trial for subsequent retention tests and as a retention test for previous acquisition trials; and, second, because the peanut happens to be the recall target in the memory task as well as the inducement to perform the task correctly. Viewed in the context of monkeys' willingness to learn performance rules defined in terms of any arbitrarily chosen histories (table 1), the development of o.d.l.s. may most plausibly be assumed to reflect the acquisition of a memory-dependent performance rule: the monkey learns to choose whichever object is associated in memory with reward.

Serial visual reversal set is not shown in table 1 and differs from the tasks in the table in that only two stimuli are used throughout, in place of the large populations of stimuli used in the tasks described so far. So in reversal set, odd-numbered reversals will consist of repeated presentations of A vs B with A rewarded while even-numbered reversals will consist of A vs B with B rewarded. But the same performance rule as in o.d.l.s. is adequate for serial reversal set and, not surprisingly, monkeys transfer readily between these two tasks (Schrier 1966; Warren 1966).

Several implications of this type of approach may be noted at this stage and are taken up in greater detail in what follows. The first is an entirely open-minded attitude to the behavioural effects of memories. Formally speaking, one may say that a memory in the current treatment is nothing more than a property of the internal state elicited by retrieval cues at retention tests. If A and B are both paired with X and C and D are both paired with Y, then the memory of X is something that A and B but not C and D will elicit if they are subsequently presented. If the animal at retention tests discriminates or can be taught to discriminate between A and B and stimuli like them on the one hand, and C and D and stimuli like them on the other, then the hypothesis that the animal remembers X, or to be more precise that the animal can remember differentially X and Y, is justified since the sole purpose of speaking in this case of memories is to refer to the shared features of stimuli whose only systematic similarity to each other is their history. There is, for example, no further implication, unless explicitly stated, that the animal should behave as if expecting X to follow A. Second, it is essential to this approach that memories are clearly distinguished from habits. The phrase 'associative learning' is often used loosely to refer to the acquisition both of memories and of habits, since, in practice, the statement that an animal has associated A with X often means only that as the experimenter sees it, the animal has developed an appropriate response tendency to A which signals X. This usage must be eschewed if the relation of memory to behaviour is the question at issue, as below. Third, as the examples of o.d.l.s. and serial reversal learning make clear, the analysis of memory processes may have implications for understanding tasks that do not initially appear to be formally designed for the study of memory-dependent performance rules.

2. Learning and habit

In many theories of learning there is no scope for learning about memories. For example, Pavlov's principle of stimulus substitution is roughly speaking the principle that if A is associated in memory with B then the animal will react to A as if it were B. We have then on the one hand an associative memory mechanism that records the association of A with B, and, on the other hand, a performance rule according to which the output of memory (recall of B, in the example) elicits behaviour; and the performance rule is not itself a product of learning but a pre-existent and fixed condition of the manifestation of learning in behaviour. Figure 1*a* is an attempt to present these relations schematically.

The account above neglects Pavlov's treatment of 'trace conditioning'. However, as Revusky & Garcia (1970) have well argued, the non-associative process that bridged the delay in trace-conditioning was not explained as part of Pavlov's theory but was added on to the main body of theory by an unexplained assumption. The word 'trace' was used in the same way by Hull (1952), to refer to an unexplained short-lasting persistence of stimuli from one trial to the next. It is of historical interest that, during the era when memories appeared in theories only in this impoverished form, experiments on memory tasks other than discrimination

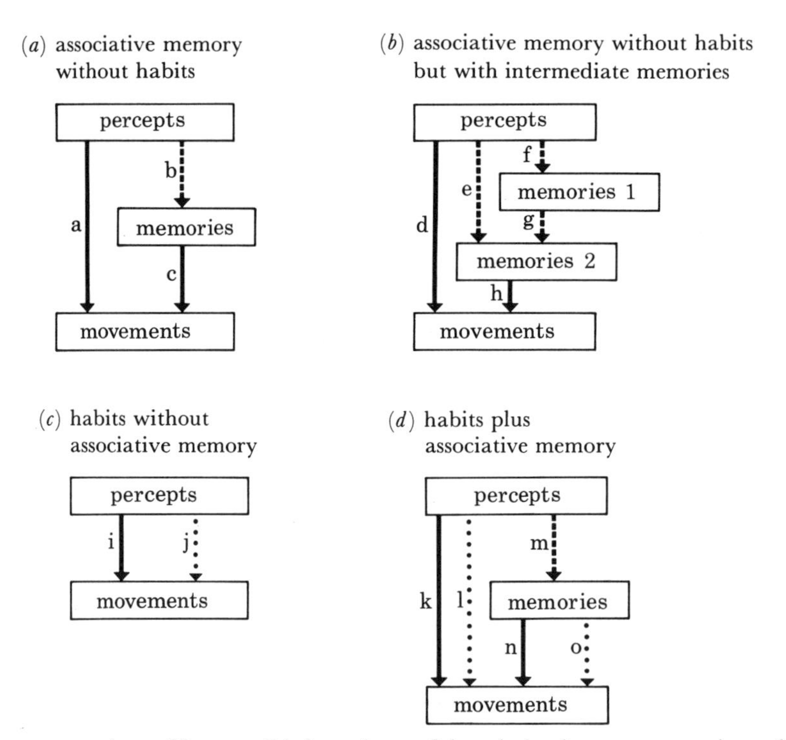

FIGURE 1. Schematic representations of four possible hypotheses of the relation between memories and movements. Solid lines indicate unlearned direct influences upon movement. Dotted lines indicate habits, namely, learned direct influences upon movement. Dashed lines indicate memory traces, that is, the ability of a retrieval cue to elicit a memory. Note that memories are treated as potentially able to do everything that percepts can do: act as a retrieval cue for a memory (line g), exert inflexible effects (line c), or elicit habits (line o).

learning required monkeys to remember only from one trial to the next. For example in delayed matching or non-matching, Nissen *et al.* (1948) and subsequently many others gave their monkeys only one sample to remember at a time, with an immediate retention test; this undemanding method may be contrasted with the lists of samples presented by Gaffan (1974), Mishkin (1978) and other modern experimentalists. Whether the lack of data held back the theories, or the lack of theory the experiments, is not clear; but at any rate it is worth stressing now that when monkeys are given lists of acquisition trials with different stimuli, their memory for events other than the reward, for example in delayed matching their memory of having seen an object, may be no less powerful than their memory for whether an object was rewarded or not (Gaffan 1976); and further that though the retention tests are presented next to the relevant acquisition trials in table 1 so that the rules should be clear to the reader, monkeys may not only perform but also learn such rules even when retention tests never follow immediately upon their objects' acquisition trials (Gaffan *et al.* 1984*c*).

With the minor exception of trace conditioning, therefore, the scheme presented in figure 1*a* is an adequate representation of the relation between associations and behaviour in Pavlovian theory. It is clearly inadequate as a model of monkeys' learning but the necessary extension of the simple model in figure 1*a* is not difficult to envisage and is presented in figure 1*b*. This model still appeals to no other process than associative memory in explaining learned modifications of behaviour, but it accommodates memories about memories. Consider, for example, matching or non-matching to sample. According to figure 1*b*, any stimulus that is associated in memory with food (line e) operates via a fixed translation rule (line h) to elicit approach. However there are, in addition, some other memories, 'memories 1' in the figure, that have no fixed unlearned effect (or only a weak fixed effect) on approach, but that can themselves act as a retrieval cue for the memory of food. Thus to take as an example the memory of having seen an object before or the memory of not having seen an object before, these memories will themselves become associated in memory with food, if it is the case that familiar or novel stimuli respectively are always rewarded with food. Then in, say, matching to sample, a sample at the retention test will elicit via line f a memory that via line g has been associated, over many similar retention tests with different objects, with food; and the sample at the retention test in matching will through this indirect process elicit approach.

Model 1*b* is a realistic proposal of at least one way that an animal could learn matching or non-matching. The idea that novelty or familiarity, as memory-retrieved properties of objects in general, are associated with food reward in matching or non-matching, receives strong support from experiments investigating the role of the sample bait at acquisition (Gaffan *et al.* 1984*d*): the sample when it appears at the acquisition trial is a novel object, and matching is therefore facilitated if the sample at acquisition, like the non-sample at the retention test, is not rewarded. Furthermore, this model is not restricted to a Pavlovian assumption of stimulus substitution. This type of account, where some associative memories have an unlearned effect upon behaviour via a fixed translation rule, includes also, for example, the inferential neo-Tolmanist aproach of Dickinson (1980, p. 115) and Mackintosh (1983, p. 111). According to this inferential hypothesis or model, unlearned logic compels a hungry rat to derive from knowledge of its hunger, and from the memory that approaching A has resulted in food, the practical inference that A should be approached, and some further unlearned process puts the practical inference into effect. In addition I may mention the version of this type of hypothesis that I myself favour.

I would suppose that just as certain percepts evoke unconditioned responses whose form is various since they are specialized to be appropriate to the individual stimuli in question, so also certain memories evoke unconditioned responses that are specially appropriate to those memories. These responses are not necessarily identical to the unconditioned responses evoked by the percepts that the memories are memories of: the memory of pain elicits responses that are not all identical with responses elicited by pain itself, and the memory of warmth elicits gentle pecking in chicks, which do not respond to warmth itself by pecking (Wasserman *et al.* 1975). This version of the hypothesis has the advantage over the Dickinson–Mackintosh version that the chicks need neither logical reasoning power nor the propositional knowledge that heat sources are likely to be hens and that the way to cajole a hen is to peck it gently. If memories are in this way seen as similar to percepts in their elicitation of unconditioned responses it becomes all the easier to see them as similar to percepts also in their potential elicitation of conditioned responses by a further association in memory. In the case of memory of peanuts in monkeys this hypothesis would however, as is often the case, be indistinguishable from the hypotheses either of stimulus substitution or of inference since all would agree that the effect is to elicit displacement of the object that elicits, either directly or via a mediating association, the memory of the food reward.

Finally in support of the generality of model 1*b*, the labelling of memories 1 and memories 2 in the figure is an aid to exposition when only one response is being considered, but a more accurate statement would be that memories can act as retrieval cues for further memories in a non-hierarchical fashion: if a monkey can learn, say, that the memory of warmth is associated with food then the animal can also presumably learn that the memory of food is associated with warmth.

However, in spite of all these strengths model 1 *b* suffers from an obvious weakness: it cannot cope with a task such as incongruent recall. A situation where the memory of food is associated with no food, and the memory of no food with food, and where the animal's behaviour is directly elicited by the memories of food and no food, is every bit as confusing as it sounds for 1 *b*. Some rear-guard action in defence of the model might have been possible if incongruent recall had been an exceptionally difficult task; unfortunately the monkeys do not find it confusing at all. In comparison with congruent recall, incongruent recall is neither a more difficult rule to learn nor a more difficult task to perform (see figure 2 and the acquisition data in experiments 3 and 4 of Gaffan (1979) and in experiment 1 of Gaffan *et al.* 1984*a*).

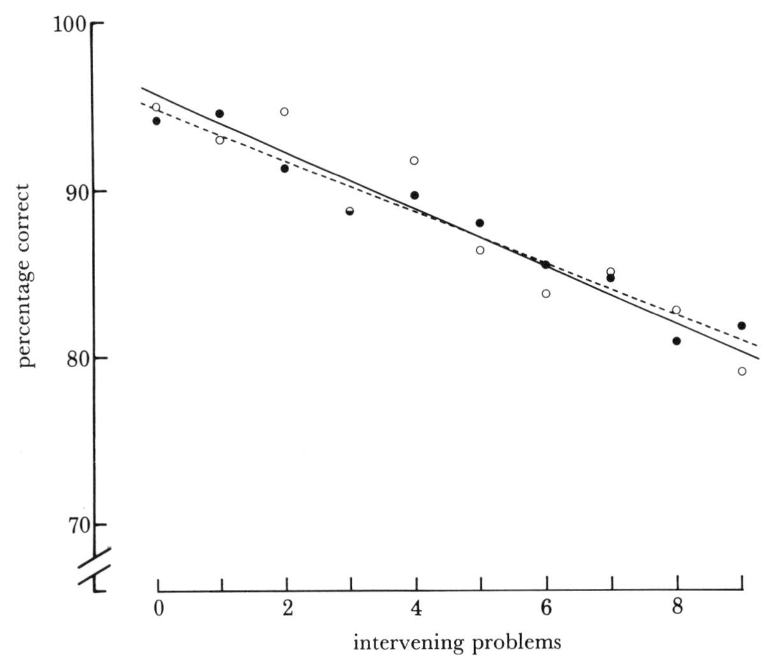

FIGURE 2. Performance of congruent recall (dashed line, filled circles) and of incongruent recall (solid line, open circles) by two groups of well-practised rhesus monkeys, containing eight and seven animals respectively. The groups are indistinguishable both in forgetting rate and in overall accuracy.

At this point it is worth stressing again the common feature of the hypotheses we have considered so far. They all rely for their explanation of learning upon the animal's acquisition of associative memories. Apart from the acquisition of associations in memory, the other mechanisms that they invoke to explain behaviour (inference, stimulus substitution, unconditioned reflexes to memories, and so on) are not themselves the product of learning; the only direct effect of learning that they allow is associative memory, and anything like a habit or a performance rule has to be explained in terms of associative memories and their sometimes complicated effects and interactions. But if 1 *a* is, with the minor exception of the impoverished 'trace' concept, an adequate depiction of Pavlov's views on the relations we are interested in, then figure 1 *c* is, with the same minor reservation, an adequate depiction of Hull's. Each of these extreme views is equally obviously inadequate to explain learning about memories, but it is nevertheless instructive to consider the hypothesis that habits unmediated by memories are one of the direct products of learning.

Consider next and finally, therefore, the composite model presented in figure 1*d*. This denies none of the mechanisms we have discussed. (The omission of the possibility of memories entering into further associations in memory is made for the sake of comprehensibility only.) Associative memories are formed and can have unconditioned effects; as well, habits are formed and can link either percepts or memories directly to behaviour. The main advantage of this model, from the evidence we have considered so far, is its ability to cope with incongruent recall. The habits envisaged in figure 1*d* are 'blind' habits in the sense that, unlike habits in the exclusively mnemonic models considered above, their performance does not require the mediation of a memory of the reward that established the habit. The habit of choosing stimuli that are associated in memory with non-reward in preference to stimuli associated in memory with reward can therefore proceed without any confusion or interference between the recall task and the motive for it.

However, mention of the motive and the reward calls attention to what the model is lacking, namely a mechanism of reinforcement. For Hull or Thorndike, reinforcement had an unalterable effect of strengthening the S–R conjunction which preceded it, and if that hypothesis was adopted then figure 1*d* would have a difficulty of its own in incongruent recall, because the peanut at acquisition would strengthen the habit of displacing that stimulus object under which it was found and would therefore detract from correct performance at the subsequent retention test. What is needed therefore is some more subtle hypothesis of reinforcement if figure 1*d* is to be viable. This is clearly not an insurmountable obstacle but before tackling it the evidence from fornix transection should be examined.

3. Impaired habit formation after fornix transection

In clinical neuropsychology several well-known and ancient lines of evidence suggest that the hippocampal system (hippocampus, fornix and mammillary bodies) is specially important for memory. In experimental neuropsychology and neurophysiology much effort over the last 30 years has gone into the attempt to provide firm evidence for the role of the hippocampus in memory and learning. Anyone who has dipped into this literature will not need to be told that in the space available here I do not intend to try to review or summarize it. However, there are some predecessors in this field of study whose influence I would particularly like to acknowledge. The remarkable correlation between hippocampal electrical activity and voluntary movement was observed and documented by Vanderwolf (1971) and Black (1975). Vanderwolf pointed out that one way to define voluntary movement is as movement whose elicitation is modified by its consequences, and that a connection might therefore be found between his observations and the behavioural evidence of some kind of learning impairment after hippocampal lesions. The study of fornix transection in monkeys was initiated by Mahut (1972). The fornix of a monkey is an accessible and clearly demarcated structure surrounded by ventricle. It is possible to transect the fornix reliably, in every animal in a group, and completely. If extra-fornical damage is produced there is no difficulty in describing the extent of that damage since the adjacent structures are clearly differentiated from the fornix, and though extra-fornical damage needs to be only slight and asymptomatic it can, if desired, be replicated in a control group. If psychologists cannot make sense of the behavioural effects of fornix transection then we may as well abandon the hypothesis that anatomically discrete structures have discrete functions that can be delineated by behavioural experimentation, since the fornix is the clearest possible example of an anatomically discrete structure and it is also by good fortune a pathway that can be cleanly and completely transected. Finally the origin of many of the ideas expressed below can be found in a lucid and strongly argued paper by Douglas (1967).

A lesion that impairs performance of a memory task may do so either by impairing the memories that are necessary for performance, or by impairing the execution of the performance rule that translates those memories into correct choices. With one important exception to be discussed below, that of incongruent recall, every task in table 1 has been shown to be performed normally by fornix-transected monkeys when they are given ample opportunity for practice at it (Gaffan *et al.* 1984*a*, *c*, *d*). The same applies to visual reversal set (Mahut 1972; Gaffan & Harrison 1984*b*) and to visual–visual memories and their reversal in a complicated paradigm to test acquisition of object–object associations (Gaffan & Bolton 1983; Gaffan *et al.* 1984*a*, experiment 3). These results suggest, therefore, that memories of the type required by these tasks are not impaired by fornix transection, and that in those instances where an impairment has been observed in tasks of this type (when, as we shall see in a moment, a new procedure was introduced after initial acquisition of the task, and performance in the new procedure was assessed without extensive practice in it) the memories were available to the monkeys but the performance rule was impaired. Since, according to model 1*d*, memory-dependent habits are similar in principle to perceptually based habits, the results from these memory tasks may imply a general impairment of some kind in the formation or modification of habits. As the following discussion shows, the evidence is consistent with that implication.

In matching or non-matching, for example, fornix-transected monkeys can show the same high levels of performance as normal monkeys (Gaffan *et al.* 1984*c*, *d*). When an impairment has been observed in these tasks it was when an established performance rule was required to be modified in response to a change in procedure (Gaffan 1974; Gaffan *et al.* 1984*c*). In the first experiment on matching and fornix transection (tasks 1 and 2 of Gaffan 1974) I began by adopting the procedure which as noted above (section 2) had been normal since the report by Nissen *et al.* (1948): the monkeys were initially trained with all retention tests following immediately upon their sample's acquisition trial. To my disappointment, the fornix-transected animals learned delayed matching just as fast as the controls. Then to test the power of the animals' memories I began introducing some very long retention intervals, and a deficit was immediately apparent in the fornix-transected group. At the time it seemed obvious that the deficiency was in long-term memory as opposed to immediate memory. It was not until several years and hypotheses later that my colleagues and I considered the possibility that the cause of this deficit and of a similar deficit in non-matching (Owen & Butler 1981) might have been the change from the conditions of original learning rather than the retention interval as such. We therefore taught non-matching to naive monkeys with a variety of retention intervals from the beginning of training: there was now no sign of an impairment in the fornix-transected group at any retention interval, although an impairment could be produced by some subsequent changes of procedure (Gaffan *et al.* 1984*c*).

This deficit in the ability to modify learned memory-based habits is seen also in perceptual habits. Fornix-transected monkeys were slow to reverse their performance in a task I shall call 'f.r.-dro reversal', where fixed-ratio and 'dro' response requirements were alternated between sessions: in the same apparatus and with the same stimuli every day, repeated manual contact (f.r.) with a stimulus was rewarded on one day and withdrawal of the hand from it (dro) was rewarded on the next (Gaffan & Harrison 1984*b*, experiment 2). In search of a contrast with serial visual reversal set which as noted above is performed normally by monkeys with lesions of the hippocampal system, we taught our monkeys a serial reversal set in this f.r.-dro reversal task. After several reversals the normal animals all reached a pre-established criterion of rapid reversal learning. None of the fornix-transected animals reached the criterion, and at the end of the experiment their within-reversal learning was still slower than the normal monkeys' even though by then they had had substantially more practice than the controls at the serial reversal task.

The modification in the animal's learned movements required from day to day in serial f.r.-dro reversal is quite different from that required in serial visual reversal learning. In serial visual reversals (as in o.d.l.s. or congruent recall) a memory is available to guide the animal's displacement of one stimulus object rather than the other. That memory, namely the memory of the food reward that sometimes one and sometimes the other object recalls, is a valid guide throughout the task. Objectively speaking, the animal's response in displacing an object switches each day from one object to the other, but as we have seen above the memory-dependent habit is not changed, only the association in memory (recall the strict distinction between habit and memory that was drawn above). But there is no similar memory to mediate between stimulus and response in the f.r.-dro reversal, since an association in memory between the stimulus features of the task and the reward outcomes is not by itself in f.r.-dro a valid guide for the animal's response. The habit itself, the tendency to make either an f.r. or a dro response in the context of the task, must change. A similar analysis applies to reversals of a simultaneous spatial discrimination, when an animal learns to go left one day and to go right the next. Fornix transection impairs performance of this task (Mahut 1972), and since the impairment is not the result of an impaired ability to discriminate spatial location as a perceptual cue (Gaffan *et al.* 1984*a*, experiment 6) it is attributable to a defect in the control of the locomotor movement, comparable to the defect in the control of manual movements that is revealed in the f.r.-dro reversal set.

Even within the paradigm of visual reversal learning it is possible to distinguish the case of habit reversal from the case of memory reversal. Within the very first visual discrimination an animal learns, say A positive vs B, A is, of course, no less valid a guide for the animal's response than the memory of reward is, and the animal might therefore acquire a perceptual habit of approaching A as well as a memory of A's association with reward. Subsequent reversal of that discrimination would then require a change in habit. In conformity with this suggestion, fornix transection did produce an impairment in the first post-operative visual reversal (Gaffan & Harrison 1984*b*, experiment 1).

In the examples discussed so far fornix transection impaired the ability to change an established habit. Some further observations suggest that this impairment can be subsumed under a more general defect, namely impaired learning ability when one habit is to be formed in one set of circumstances and a different habit is to be formed in a different set of circumstances that is similar to the first and therefore liable to be confused with it. In reversal learning the two circumstances follow each other, but in learning a conditional response they are intermingled in time and the formation of the two habits proceeds simultaneously with neither necessarily taking precedence. An example of such conditional response learning is a task where the animal must go left in the presence of one set of visual stimuli and go right in the presence of a different set. This task was learned abnormally slowly by fornix-transected monkeys (Gaffan *et al.* 1984*a*, experiment 5). We have observed a similar impairment in recent unpublished experiments with visual conditional learning of the f.r. and dro responses.

In reversal learning the contrast between visual reversal set and response reversal set shows that the lesion effect is concerned with some aspect of the learned control of the animal's own movements, since if there had been a general impairment in reversal as such there would have been a deterioration in the rate of reversal of memories in the visual reversal set. The same point can be made about impaired conditional response learning. Configural discriminations, which involve conditional relations between stimulus properties rather than between stimuli and the animal's own movements, were learned normally. There was no impairment in learning an auditory–visual configural task where an auditory cue signalled which of two visual stimuli was rewarded, and there was also no impairment in learning a visual–spatial configural task (Gaffan *et al.* 1984*a*, experiments 6 and 7). Thus the impairment in conditional response learning is not a general impairment in learning about conditional relations.

Finally, consider one more example of impaired habit formation, in this case of a memory-dependent habit. The explanation offered above for normal animals' high levels of performance in incongruent recall was that they formed a memory-dependent habit. This habit requires some reversal learning in the sense that monkeys that are experimentally naive show some tendency to perform according to the performance rule of congruent recall (Gaffan 1979, experiment 3); it also requires a kind of conditional response to be learned, since memories similar to and liable to be confused with those that control incongruent recall performance control something more like congruent recall performance in the animal's daily experience of extra-experimental feeding. As might therefore be expected, fornix transection severely disrupted incongruent recall performance (Gaffan *et al.* 1984*a*, experiment 1).

In conclusion then fornix transection impairs habit formation both in line l and in line o of figure 1*d*. It does not completely abolish it: as the examples illustrated have shown, fornix transected animals form habits more slowly, but they do form them. The contrast is with sensory memories where the memories are acquired without impairment at a normal rate of acquisition. The deficit in habit formation is especially evident when there is the possibility of confusion between conflicting habits appropriate to similar circumstances, as in reversal or conditional learning. These two features, the survival of some capacity for habit formation, and the special difficulty in discriminating habits that are liable to confusion, may not be theoretically distinct since it is presumably in the face of such liability that efficient acquisition of appropriate habits is most necessary. But the idea of a perceptual stimulus eliciting a voluntary movement via a habit requires careful definition. Fornix transection does not produce a general difficulty in acquisition of reversed or conditional memories, actions or percepts, as is shown by the examples above of reversal in sensory memories and of learned performance rules that involve conditional relations between percepts; the difficulty observed is therefore something specially to do with the animal's own learned movements as opposed to events in the environment; and yet there is no impairment in the smooth execution of movements, nor in postural adjustment, nor in the sequencing and control, presumably via sensory feedback, of a complex behaviour such as the emission of an f.r. response. We may think therefore of a habit as a mechanism for conveying a vague instruction ('leave it alone', 'go left', 'hammer it') from the visual or auditory modalities, in which the cue for the habit is recognized, to a motor command that has its own complex routines to determine in detail just how a dro response, a locomotor response or an f.r. response (respectively) is constituted in terms of muscular contraction and sensory feedback. The function of the fornix, then, is to facilitate the acquisition of excitatory links which carry vague instructions from the temporal lobe where perceptual recognition takes place to the thalamus where movements are initiated. But the facilitation of the acquisition of habits is precisely what is meant by reinforcement, of which as noted at the end of the previous section a hypothesis is required. Not surprisingly then, further examination of effects of fornix transection may supply such a hypothesis.

4. Reinforcement, exploration and response memory

Exploration in animals is by no means a matter of running around aimlessly when there is nothing better to do. It is a memory-dependent performance rule, though presumably an innate one. Exploration is directed towards alternatives that are novel in the sense that they have not recently been explored, and a memory of past behaviour is therefore required in exploration to direct present behaviour towards novel choices. Much research upon exploration is performed with rats and is concerned with their locomotor choices between spatially defined alternatives, but exploration can be similarly demonstrated in monkeys' choices between visually defined alternatives. Naive monkeys were tested in a paradigm identical with matching or non-matching to sample except that both objects at the retention test were rewarded. Since the sample at acquisition was also rewarded this was a pleasant task in which every displaced object revealed a peanut. Normal monkeys showed a tendency towards spontaneous non-matching. That tendency was abolished by fornix transection (Gaffan *et al.* 1984*c*).

Consider therefore the hypothesis that since exploration is directed by memory, and since damage to the hippocampal system impairs exploration in animals and memory in men, the direct effect of the fornix transection in the experiment above might be to impair memory in some way and thus to cause indirectly the observable decrement in exploration. If this hypothesis is true then the memory impairment in question must affect some aspect of memory that is more important for spontaneous non-matching than for trained non-matching, since the latter is unimpaired. The obvious possibility was that the crucial aspect of memory might be the animal's memory for its own responses to stimuli as opposed to the sensory familiarity of the stimuli in themselves, since the function of exploration is to direct the animal's movements. This possibility was easy to test. A monkey will displace an object that covers a food well, but will leave alone an object that is just behind the food well, even though the latter object will be seen if the food well is baited and a peanut is retrieved from it. To return to the terminology of § 1, we have here a difference between two histories. If an object is placed on top of a baited well then its history will include the fact that the monkey displaced it. If an object is placed instead just behind a baited well then its history will be similar to that of the first object except that it will not have been displaced. So one can set up a memory task to distinguish between these two histories, either by rewarding, at subsequent retention tests, objects that have been displaced and not rewarding objects that have not been displaced, or vice versa. The task may be called respectively 'push-matching' or 'push-non-matching'. Both of these tasks were impaired by fornix transection and this impairment was not alleviated by practice (Gaffan *et al.* 1984*c*, *d*). There was thus a clear contrast between memory for responses and memory for the various aspects of environmental events that are involved in the unimpaired tasks of table 1.

The distinction between memory and habit is once again crucially important. When a stimulus elicits a memory of the animal's previous response to that stimulus it is difficult not to think in terms of an 'S–R association', but the connotation of that term is entirely inappropriate to the present context. In the terms of the conclusion of §3 above, the memory of a previous response is not the issuing of a vague instruction but instead the memory of a vague instruction having been obeyed. It is only if this distinction is clear that one can ask the interesting question: what is the relation between the memories of previous responses to itself that a stimulus evokes, and the habit that that stimulus elicits?

A concise answer to that question is that reinforcement is the converse of exploration: a habit is what is left after exploration has ceased. Moreover, having defined exploration as the choice of a novel response, which is already a memory-dependent performance, one can further propose that at a higher level exploration itself is controlled by memories. The form of control would be that, other things being equal, if a stimulus is familiar, and evokes a memory of consistent previous responses to itself, and also evokes no memory of any recent surprises, then that stimulus suppresses exploration in favour of the opposite performance rule to exploration, namely habit, the production of the response that the stimulus recalls.

We may now return to the question that was posed at the end of §2. It can now be seen that the real problem of the cognitive model of habit formation was not its appeal to associative memories but more specifically its appeal to memory of reward. The hypothesis that habit is to be explained by unconditioned responses to memories has no difficulty in allowing for the habitual choice of stimuli associated with non-reward in incongruent recall, so long as the memories that in turn elicit the performance of the habit are not themselves memories of reward but, as envisaged above, memories of the previous performances of the habit.

The fact that the cognitive model of reinforcement appeals to memory of the reward that established the habit is no coincidence. In the everyday language which cognitive psychology adopts, memory is in the paradigm case the ability to report verbally upon the past, and intentional or voluntary action implies an ability to justify the action verbally. In the case of a reinforcement as used to induce learning of a task, that justification would of course be stated in terms of the past delivery of the reinforcer. To make the argument from the other point of view, a cognitive psychologist would have every reason to object to the hypothesis formulated above if it were offered as a piece of cognitive psychology, that is as an explanation of habitual action in terms of knowledge. A man may even perform a habit to find out what he habitually does in a certain circumstance, because he cannot remember what he does; more generally, it would be difficult to claim that every habitual action is preceded by a memory of itself in the cognitive sense of memory as conscious awareness of the past. The technical sense of memory defined above (§1) is, of course, another matter. The study of physiological mechanisms of behaviour leads one by gradual steps away from the natural-language categories in which questions about behaviour are most easily formulated in the initial stages of that study.

References

Black, A. H. 1975 Hippocampal electrical activity and behavior. In *The hippocampus*, vol. 1 (ed. R. L. Isaacson & K. H. Pribram), pp. 129–167. New York: Plenum.

Dickinson, A. 1980 *Contemporary animal learning theory*. Cambridge: University Press.

Douglas, R. J. 1967 The hippocampus and behavior. *Psychol. Bull.* **67**, 416–442.

Gaffan, D. 1974 Recognition impaired and association intact in the memory of monkeys after transection of the fornix. *J. comp. physiol. Psychol.* **86**, 1100–1109.

Gaffan, D. 1976 Recognition memory in animals. In *Recall and recognition* (ed. J. Brown), pp. 229–242. London: Wiley.

Gaffan, D. 1979 Acquisition and forgetting in monkeys' memory of informational object-reward associations. *Learning Motiv.* **10**, 419–444.

Gaffan, D. & Bolton, J. 1983 Learning of object–object associations by monkeys. *Q. Jl exp. Psychol.* **35 B**, 149–155.

Gaffan, D., Saunders, R. C., Gaffan, E. A., Harrison, S., Shields, C. & Owen, M. J. 1984*a* Effects of fornix transection upon associative memory in monkeys: role of the hippocampus in learned action. *Q. Jl exp. Psychol.* **36 B**, 173–221.

Gaffan, D. & Harrison, S. 1984*b* Reversal learning by fornix-transected monkeys. *Q. Jl exp. Psychol.* **36 B**, 223–234.

Gaffan, D., Gaffan, E. A. & Harrison, S. 1984*c* Effects of fornix transection upon spontaneous and trained non-matching by monkeys. *Q. Jl exp. Psychol.* **36 B**. (In the press.)

Gaffan, D., Shields, C. & Harrison, S. 1984*d* Delayed matching by fornix-transected monkeys: the sample, the push and the bait. *Q. Jl exp. Psychol.* **36 B**. (In the press.)

Hull, C. L. 1952 *A behavior system*. New Haven: Yale University Press.

Mackintosh, N. J. 1983 *Conditioning and associative learning*. Oxford: University Press.

Mahut, H. 1972 A selective spatial deficit in monkeys after transection of the fornix. *Neuropsychologia* **10**, 65–74.

Mishkin, M. 1978 Memory in monkeys severely impaired by combined but not by separate removal of amygdala and hippocampus. *Nature, Lond.* **273**, 297–298.

Nissen, H. W., Blum, J. S. & Blum, R. A. 1948 Analysis of matching behavior in chimpanzee. *J. comp. physiol. Psychol.* **41**, 62–74.

Owen, M. J. & Butler, S. R. 1981 Amnesia after transection of the fornix in monkeys: long-term memory impaired, short-term memory intact. *Behav. Brain Res.* **3**, 115–123.

Revusky, S. & Garcia, J. 1970 Learned associations over long delays. In *The psychology of learning and motivation*, vol. 4 (ed. G. H. Bower), pp. 1–84. New York: Academic Press.

Schrier, A. M. 1966 Transfer by macaque monkeys between learning-set and repeated-reversal tasks. *Percept. Motor Skills* **23**, 787–792.

Vanderwolf, C. H. 1971 Limbic–diencephalic mechanisms of voluntary movement. *Psychol. Rev.* **78**, 83–113.

Warren, J. M. 1966 Reversal learning and the formation of learning sets by cats and rhesus monkeys. *J. comp. physiol. Psychol.* **61**, 421–428.

Wasserman, E. A., Hunter, N. B., Gutowski, K. A. & Bader, S. A. 1975 Autoshaping chicks with heat reinforcement: the role of stimulus–reinforcer and response–reinforcer relations. *J. exp. Psychol. Anim. Behav. Processes* **1**, 158–169.

Phil. Trans. R. Soc. Lond. B **308**, 101–111 (1985) [101]
Printed in Great Britain

Cortical mechanisms and cues for action

By R. E. Passingham
Department of Experimental Psychology, South Parks Road, Oxford OX1 3UD, U.K.

Monkeys have more highly developed brains and are more intelligent than rats; yet rats learn some tasks as efficiently as monkeys. For example, rats are as quick at discovering which of two doors hides food or how to open the doors. Presumably tasks of this sort do not greatly tax cortical associative mechanisms since the animals have only to cumulate facts about objects. It is argued that cortical mechanisms are crucial for the ability to relate together information that is presented at different times or in different places. After removal of parts of frontal cortex monkeys can still associate cues that are presented together but they are poor at relating cues that are presented apart.

Compared with monkeys rats have very underdeveloped brains. Even if a rat were as large as a rhesus monkey its brain would still be 3.2 times smaller (calculated from Eisenberg 1981). In the laboratory rat the neocortex (with the white matter) forms only 29.7% of the whole brain, whereas in a rhesus monkey the value is as high as 72.2% (calculated from Kruska (1975) and Stephan *et al.* (1981)). It would be reasonable to expect that the rat would lag considerably behind a monkey on tests of cognitive performance.

Yet on certain tests rats rival monkeys. On a visual discrimination task the animal must learn to associate one cue with food and the other with no food. In our laboratory both rats and rhesus monkeys (*Macaca mulatta*) have been given a simultaneous visual discrimination between black and white: the rats pushed doors on which the cue was displayed and the monkeys pushed plastic covers. Though the monkeys tended to learn faster there were rats that out-ranked some of the monkeys (figure 1).

The demands of the task can be increased by requiring the animal to perform one act to one cue and a different act to the other cue. We have taught a conditional problem of this sort to both rats and monkeys. The rats had to open a door by pushing it if it was white and pulling it if it was black (Passingham *et al.* 1985); the monkeys (*Macaca mulatta*) had to pull a handle if it was yellow and turn it if it was blue (Passingham 1985*b*). The task was more difficult than the visual discrimination; but still the rats learnt as easily as the monkeys (figure 2).

On both tasks the animals must form associations, and on both tasks the cues are arbitrary. Yet monkeys fail to outstrip the rats. In general different vertebrate species do not differ very markedly in the rate at which they solve discrimination problems or master the use of a manipulandum (Warren 1965). Why, then, do these tasks pose so little problem to the animal? The tasks are alike in that the animal must learn a property of an object. Either the object covers food or it is to be used in a particular way. Many problems that animals face in nature are of this sort: the animal must learn which foods are edible and how they are to be handled or opened. For example, monkeys deal with nuts and bananas in a different fashion. In the laboratory monkeys have little difficulty in learning discriminations if it is the foods themselves

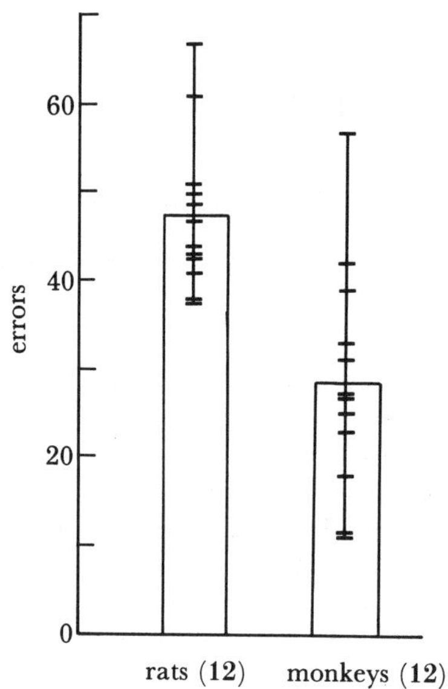

FIGURE 1. The mean errors to criterion for rats and monkeys to learn a simultaneous discrimination between black (positive) and white (negative). The bars give the scores for the individual rats or monkeys.

that they are given to discriminate or if the cue directly hides the food (Jarvik 1953; Cowey 1968). The animal's task is simply to remember information that is presented together in one place.

Tasks of this sort make few demands on higher cognitive mechanisms. It is usually supposed that one measure of higher intelligence is the capacity to grasp relationships (Thomas 1980). Thus the animal's ability is best tested not by asking the animal to learn that one object hides the food but by asking it to compare two objects and judge whether they are the same or different (King & Fobes 1982). The conditional problem can also be made more demanding. The animals in figure 2 were learning how to use an object on the basis of a property of that same object. The task is more taxing if the cue is separate from the object and the animals must therefore relate the nature of the cue to the usage of the object.

We have devised two tasks of this sort for monkeys. As before the animal has to pull or turn a handle, but the handle always has the same appearance. In one version the instruction is given by the colour of a plastic panel in front of the handle; the monkey starts the trial by

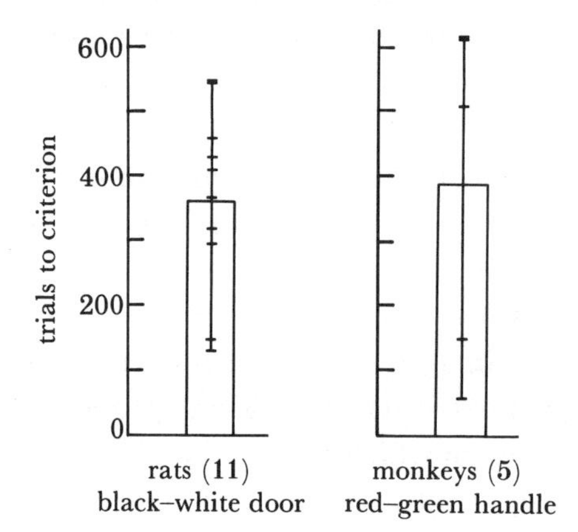

FIGURE 2. The mean trials to criterion for rats and monkeys to learn a visual conditional motor task.

pushing the panel to one side and must pull the handle if the panel was blue and turn it if the panel was yellow (Halsband & Passingham 1982). In an alternative version it is the colour of the background that provides the cue; the background is a perspex panel that is lit either blue (pull) or red (turn). The task is difficult to teach. We have trained eight monkeys (*Macaca fascicularis*) on the version with the background cue, and only four of them had passed in 1850 trials. Yet the four animals that failed had no trouble in learning to pull or turn the handle if it was the handle that gave the cue. We presented either a small or a large handle: if it was small they had to pull it, if large turn it. The animals took only a mean of 261 trials to master this problem. The animals find it easy to understand that the identity of the object is relevant to the way in which it should be handled; but they are slow to appreciate the relevance of environmental conditions or context.

1. Brain mechanisms

Cues for action

There is a way to test the claim that these tasks require different mental operations and thus differ in kind. This is to find an experimental manipulation such that animals who pass one task fail the other. One such manipulation is to tamper directly with the brain mechanisms themselves. We have been investigating the effects of removing premotor cortex (area 6) on the ability of monkeys to perform conditional motor tasks. Figure 3 shows the subdivisions of

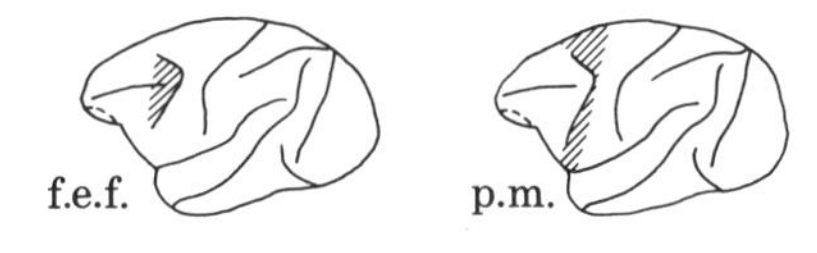

FIGURE 3. Location of lesions mentioned in the text for the macaque brain. p.m., Premotor cortex; f.e.f., frontal eye-fields; s.p., sulcus principalis; s.c., superior prefrontal convexity; i.c., inferior prefrontal convexity. The motor area lies behind premotor cortex in and in front of the central sulcus.

frontal cortex into motor cortex, premotor cortex and prefrontal cortex. After removal of premotor cortex monkeys (*Macaca mulatta*) had little or no difficulty in relearning the task if it is the colour of the handle that indicates whether it is correct to pull or turn (figure 4) (Passingham 1985*a*). Yet other monkeys (*Macaca fascicularis*) with a premotor lesion failed to relearn the task when the correct movement was signalled by the colour of the panel in front of the handle (Halsband & Passingham 1982).

As a check a further comparison was made. Instead of requiring the animals to pull or turn they were required only to touch the object, but to do so sometimes with the left hand and sometimes with the right. Monkeys can be trained to use one or other hand according to a visual cue (McGonigle & Flook 1978). Two versions of the task were used: in one they were to use the left hand to push the object if it was object A and the right if it was object B; in

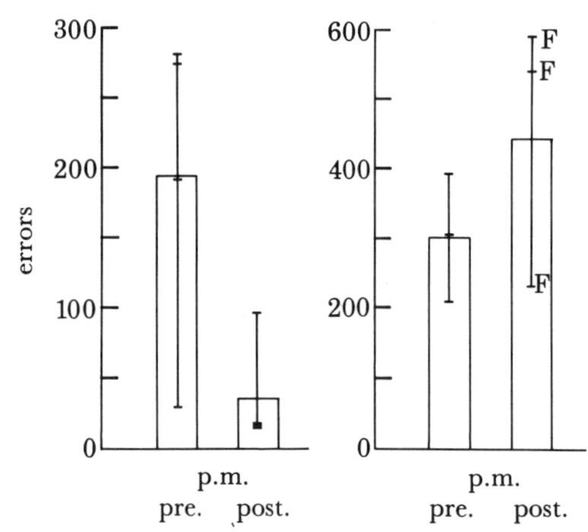

FIGURE 4. Errors to criterion for monkeys with premotor lesions (p.m.) to learn two versions of a visual conditional motor task. For figure on left cue is given by colour of handle. For figure on right cue is given by colour of panel. pre., preoperative learning; post., postoperative retention. Note that the scale for the ordinate differs for the two figures. F, failed in 1000 trials.

the other version they were to push a clear panel with one hand if the bulb behind it was lit red and with the other hand if the bulb was lit green. It was very much more difficult to teach the second than the first of these tasks (figure 5). After the removal of dorsal premotor cortex (and the frontal eye-fields) monkeys (*Macaca mulatta*) had no trouble in learning the use of the correct hand to push the objects; yet they were very poor at learning to push the panel according to the background illumination (figure 5) (Halsband 1982). They could base their action on the identity of the object being used but not on the environmental context.

This dissociation between tasks would be of little interest if it was simply that the monkeys failed to notice the cues unless they were properties of the objects being used. It is implausible that the monkeys would fail to notice the coloured panel as they had first to push it to one side, but the issue is open to test. We need to show that the animal can use the cue if given a different task. Thus Petrides (1982) has reported that after removal of premotor cortex (area

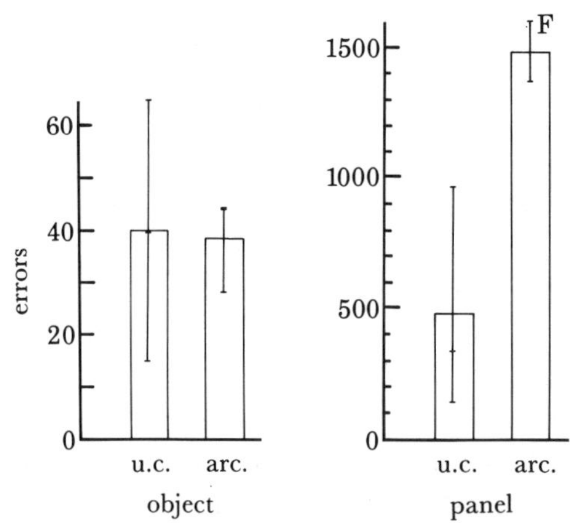

FIGURE 5. Errors to criterion for monkeys with dorsal arcuate lesions (arc.) to learn two versions of task cueing the use of the left or right hand. For figure on left cue is given by the colour and shape of the object. For figure on right cue is the background illumination. Note that the scale of the ordinate differs for the two figures. u.c., Unoperated control; F, failed in 3500 trials.

6 and area 8) monkeys fail to learn to make one of two movements depending on which of two objects appeared in the background; yet the same animals could learn whether or not to touch a rod depending on the object presented at the back. In the first case, but not the second, the object acts as a prompt as to *which* act is to be performed. The animals noticed the cue but they could only use it to tell them whether food was available or not, not what movement to perform.

There is a third way of showing that monkeys with lesions in premotor cortex can use cues to guide their actions so long as the cues are provided by the object being manipulated. This is to require that they move a handle on the basis of prior information about how the handle could be moved. We taught monkeys (*Macaca mulatta*) to reach through a hole and to squeeze or turn a handle that they were unable to see. On each trial the handle was initially locked so that only one of the two movements could be made. The animal was thus forced either to squeeze the handle by flexing the fingers or to turn it by rotating the wrist. Five seconds later it placed its hand on the handle again and was given a free choice of making either movement. The monkeys were rewarded for repeating the movement they had just been forced to make. when the animals had learnt the task dorsal premotor cortex was removed in three animals, together with the tissue in the anterior bank of the upper limb of the arcuate sulcus (Passingham 1985*b*). Although one of the three animals was impaired two of the animals relearnt with few errors; and all three animals relearnt much more quickly than before operation (figure 6). Yet the task is a difficult one; before the operation it took the six animals a mean of 675 trials (352–995) to learn the task.

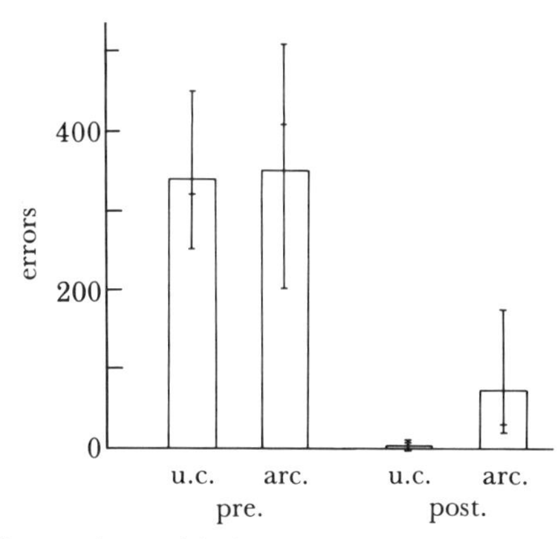

FIGURE 6. Errors to criterion for monkeys with dorsal arcuate lesions (arc.) to learn a motor conditional task in which the cue was a movement (squeezing or rotating a handle). u.c., Unoperated control; pre., preoperative learning; post., postoperative retention.

It seems that the animals understand the relation between the proper use of an object and the nature of that object but not between the use of the object and the environmental context. A similar dissociation can be demonstrated in the case of visual discrimination tasks such as the black–white discrimination discussed earlier. Mishkin *et al.* (1982) have given monkeys two tasks. On the first the animal is presented with pairs of objects and must learn which object is associated with food. Each pair is presented for just one trial a day so that 24 h separate each trial. On the second task the monkey is given an object and then 10 s later required to pick that object out again from a pair. After removal of medial temporal structures monkeys were very poor at the second task but had no trouble with the first. Yet the first task required that the animals retain what they learn over 24 h while on the second task memory was tested over 10 s.

The matching task differs in one crucial respect (see Honig 1978; Rawlins 1985). *Within* each trial the animal must relate one object (the sample) to other objects presented at a later time (the choice objects). On the discrimination task the animal does not have to put together information in this way. The same event repeats on different trials, for example food is found under one object; and the animal has only to record and cumulate the number of occasions on which this event recurs. In the experiment of Mishkin *et al.* (1982) the monkeys with medial temporal lesions could accumulate information about the properties of objects (that they were associated with food) but were poor at comparing objects presented at different times.

Thus the brain treats quite differently the task of learning about the properties of objects (association with reward or correct usage) and the task of relating one object or cue to the task of dealing with another object. In the first case the association is made between information that is presented in the same place and at the same time. The animal must simply remember that information and cumulate its knowledge over trials. In the second case the association is made between information that is presented apart. The animal must perform the cognitive operation of relating the bits of information so as to identify the current situation.

Spatial contiguity

Things and events can be related in space or time. It is possible to dissociate the cortical mechanisms that handle spatial or temporal relations. Experiments can vary the separation between cues in space or the interval between the presentation of a cue and the time at which the animal must respond.

Thus spatial contiguity can be varied by presenting a task on which the animals have the choice of opening the food well on their left or right. A visual cue, say a colour, tells them where the food is to be found. In the contiguous version of the task the cue is visible by the food wells; in the discontiguous version the cue is placed centrally and the animals must first take note of it before responding to the food wells.

Passingham (1971) taught monkeys (*Macaca mulatta*) conditional tasks in which the cue was contiguous with the place to which the animals responded. There were two tasks. In one the cue was a colour: if the board in which the food wells were located was orange the food was on the left; if blue on the right. In the other task the cue was a spatial one: the animals were given a choice between two objects, A and B; if the objects were on their left A was correct and if the objects were on their right then B was correct. In four monkeys dorsolateral frontal cortex was removed, including the frontal eye-fields (figure 3). Yet these monkeys learnt the task as quickly as unoperated animals whether the cue was colour or spatial position (figure 7). They had no trouble in associating the cue with the correct choice if the cue was clearly evident as they opened the food well.

But monkeys with lesions in the frontal eye-fields are impaired at learning to locate food if the cue is presented somewhere other than the point to which they respond. Milner *et al.* (1978) taught monkeys (*Macaca fascicularis*) to choose between left and right on the basis of a central colour cue. After the bilateral removal of the frontal eye-fields (figure 3) the monkeys were slow to relearn the task. That it is indeed the location of the cue that is crucial was clearly demonstrated by Lawler (1980). She directly compared contiguous and non-contiguous

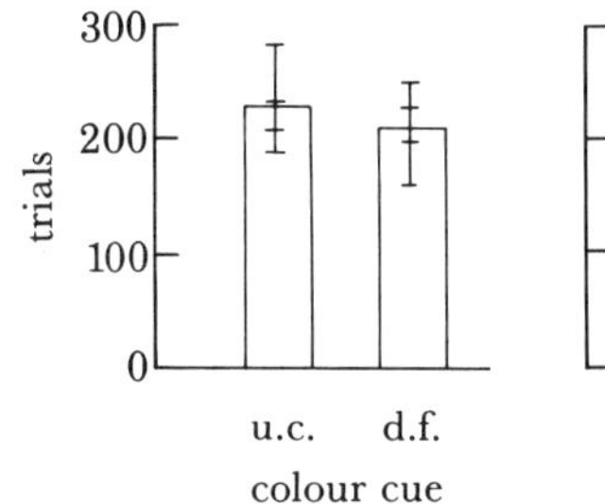

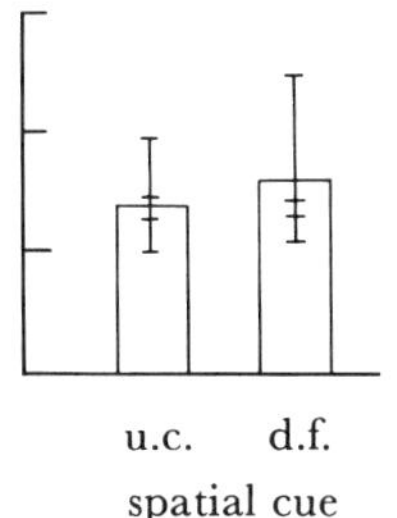

FIGURE 7. Trials to criterion for monkeys with dorsolateral frontal lesions (d.f.) to learn two versions of a conditional task. For figure on left cue is colour, for figure on right cue is spatial. For details see text. u.c., Unoperated control.

versions of the task. A cue, black or white, told the animal whether to open the left or right panels. After removal of the frontal eye-fields on both sides the monkeys (*Macaca mulatta*) learnt easily if the panels gave the cue but they were impaired if they had to use a central cue.

This would not be worthy of note if it were simply that these animals failed to notice the cue at all if it was placed centrally. But we know that the animals did notice the cue since to start the trial they had to push the centre panel to retrieve half a peanut. It is more likely that they failed to pay further attention to the cue once they looked to the left or right. After removal of the frontal eye-fields monkeys tend to neglect visual stimuli and are slow when required to search for them in an array (Latto 1982; Collin *et al.* 1982). They detect stimuli that are presented on their own, but will ignore one stimulus if at the same time another stimulus is presented elsewhere (Rizzolatti *et al.* 1983). In the human clinical literature a defect of this sort is termed an 'extinction' defect.

Temporal contiguity

Temporal contiguity can be varied by presenting the cue either at the time the animals makes its choice or in advance of this time. If the tissue is removed in sulcus principalis (figure 3) monkeys (*Macaca mulatta*) can still choose accurately between two food wells on their left and right so long as the cue is present at the time of choice (Passingham 1985*a*). A light appeared either on the upper or the lower of two central panels and the monkey started the trial by pressing the relevant panel. If the upper panel was lit the monkey had to open the left hand door, if the lower panel the right hand door. The monkeys could learn the task normally after the operation even though there was a spatial discontiguity between the spatial cue and the doors to which they responded (figure 8).

Yet monkeys with this lesion fail to learn the task at all if the cue is presented a second or two before the time at which they are allowed to respond. On the standard delayed response task the spatial cue is presented by allowing the animal to watch a peanut being placed in the food well on the left or the right. After removal of the tissue in sulcus principalis monkeys (*Macaca mulatta*) perform at chance or near chance levels (Goldman *et al.* 1971). On the delayed spatial conditional task the animals must choose between left and right on the basis of a non-spatial cue such as a colour or pattern cue, the cue no longer being present at the time of choice. If electrical stimulation is applied so as to disrupt the activity of the tissue in sulcus principalis monkeys (*Macaca mulatta*) are very poor at this task (Cohen 1970).

Damage to other areas of prefrontal cortex also impairs the ability of monkeys to respond

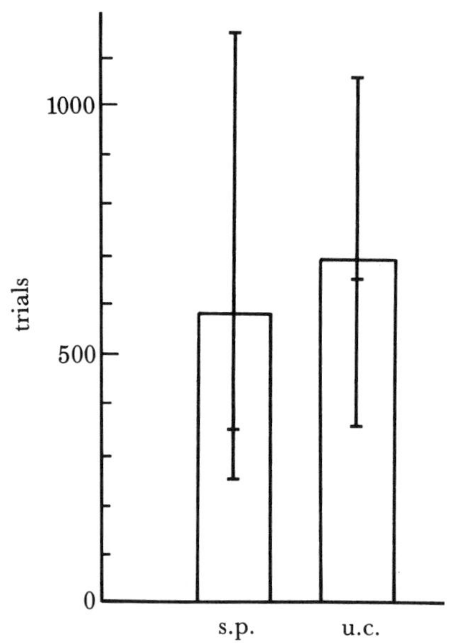

FIGURE 8. Trials to criterion for monkeys with lesions in sulcus principalis (s.p.) to learn a spatial conditional with no delay. u.c., Unoperated control.

accurately on instruction from cues that are no longer available. After removal of the tissue of the superior prefrontal convexity (figure 3) monkeys (*Macaca mulatta*) are poor at a counting task (Passingham 1978). This requires them to press a lit key and to continue pressing until the light goes off; they must then press a second key the same number of times without any further cue as to when to stop. After removal of the tissue of the inferior prefrontal convexity (figure 3) monkeys (*Macaca mulatta*) are poor at a colour matching task (Passingham 1975). On this they must press a central key which is lit either red or green; the cue goes off and they must then press the one of two other keys that bears the same colour.

On all these 'delayed' tasks the animal must make use of an instruction that they can no longer see. There is no evidence, however, that it is crucial that there be a long delay before the animal is allowed to make its choice. On the standard delayed response task monkeys with lesions in sulcus principalis can fail the task completely even if they are allowed to choose only a second or so after being shown the location of the food (Goldman *et al.* 1971). It is important only that their view of the correct location is interrupted (Kojima *et al.* 1982). Similarly both on the counting task and on the matching task impairments can be demonstrated even if the monkey is allowed to respond immediately after the cue has been withdrawn (figure 9).

Either the monkeys suffer from an impairment of memory or they fail to make use of the information even though they have stored it (Passingham 1985*a*). Consider the monkeys with lesions in sulcus principalis. They can learn a non-delayed spatial conditional (figure 6). Thus they are able to use an external spatial cue to guide their choice; and they can also learn to associate one location (left) with one cue (up) and another location (right) with the other cue (down). Yet they are unable to make sense of the problem if the spatial cue is internal even though it was visible only a second before.

One possibility must be that when they are faced with an external stimulus that demands attention they fail to attend to an internal stimulus. By analogy with visual neglect this would be equivalent to an extinction defect. As a consequence the monkeys would fail to relate internal

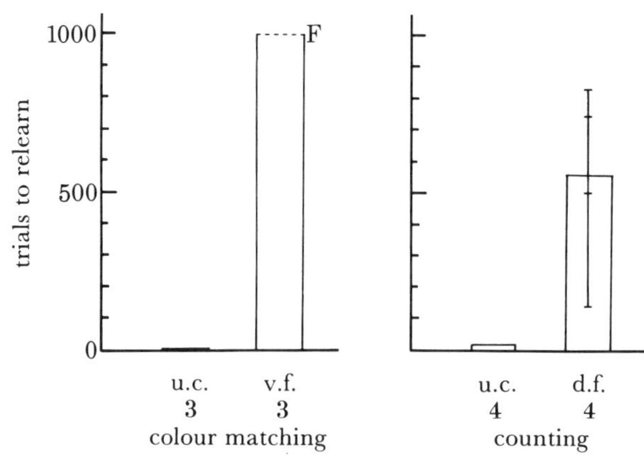

FIGURE 9. Trials for postoperative relearning of a colour matching task (left) and a counting task (right). Monkeys with lesions of the inferior prefrontal convexity (v.f.) were taught the matching task. Monkeys with lesions of the superior prefrontal convexity (d.f.) were taught the counting task. u.c., Unoperated controls.

to external stimuli. On this view prefrontal cortex plays an essential role not in memorizing *per se* but in performing cognitive operations on information stored in memory (Weiskrantz 1982).

In general frontal cortex has the task of relating cues to action. Monkeys with premotor lesions are poor at guiding their action according to visual cues unless those cues are properties of the objects they are using. Monkeys with lesions in the frontal eye-fields neglect cues that are in the periphery of their visual field when they act. Monkeys with lesions in prefrontal cortex fail to act according to internal cues. In all cases the monkeys fail to take account of relevant cues.

The claim that frontal cortex is an associative mechanism is not a new one (French 1964). But it is important to recognize that frontal mechanisms are crucial only where the items to be related are presented apart, whether in space or in time. The monkeys must then appreciate the relevance of cues that are not immediately evident as they act. To assess the current situation they must search for the relevant cues either in their environment or in their head.

2. Conclusions

Different species

These studies of frontal cortex all presented the animals with conditional tasks on which the correct action was conditional on the cue presented. Tasks of this sort are taxing for animals. Consider the delayed response problem, a spatial conditional task. This task has been given to rhesus monkeys (*Macaca mulatta*) at different stages of development (Harlow & Mears 1979). Monkeys who start training at 60 days old fail to reach a level of 80% correct trials in 900 trials, whereas adults reach 90% correct in around 150 trials. Yet if monkeys are presented with a black–white discrimination problem at the age of 11 days they can solve it rapidly (Zimmerman & Torrey 1965).

In young monkeys prefrontal cortex is not functionally mature. If prefrontal cortex is removed in a young monkey (*Macaca mulatta*) the animal can still learn the delayed response task at the age of one year (Goldman 1974). Cooling the tissue in sulcus principalis has no effect on the animal's performance at the age of 9–16 months but it has a disruptive effect in mature animals (Goldman & Alexander 1977). The reason that a lesion has no effect in an infant animal is that the prefrontal tissue is not yet making a major contribution.

Similar reasoning will account for data comparing different species, as opposed to the same species at different ages. A rat learns as quickly as a monkey to discriminate black from white (figure 1) or to guide its actions according to the nature of the object being used (figure 2). If neither task makes much demand on cortical associative mechanisms the rat is not penalized, even though these mechanisms are less well developed than in the monkey.

Cortex

On the basis of the evidence reviewed here it is possible to offer a very general account of the nature of these cortical mechanisms. Cortex is well suited to the task of relating information. It is constructed as a layered sheet and the fibres connecting different areas are gathered underneath. There is a great mass of such fibres. In a macaque monkey the neocortex on its own forms 50.2% of the total volume of brain (Harman 1947); but if the underlying white matter is included the figure rises to 72.2% (Stephan *et al.* 1981). Thus in a macaque the white matter, including incoming and outgoing fibres, accounts for around 20% of the volume of the brain. It is easiest to provide for such rich interconnections if the structure is built as a sheet like a printed circuit board. So dense a mass of connections is less easily accommodated within the structure of a nucleus.

With modern neuroanatomical techniques it has proved feasible to trace some of the interconnections between the various cortical areas. It is now known that there is a hierarchical organization: primary sensory areas project to secondary areas, and these in turn to other association areas (Jones & Powell 1970; Pandya & Seltzer 1982). Situated at the top of the hierarchy frontal cortex (premotor and prefrontal) receives information from association areas processing information in all sense modalities. The fact that information is so readily transmitted between cortical areas led Philips *et al.* (1984) to point to a 'distributive' function of cortex.

But it is not simply that information is passed between areas. The hierarchical organization allows for the bringing together of information. If the animal is to relate the elements of a situation, and if the elements are presented in different places or at different times, then the elements will not be analysed initially in exactly the same cortical region. The brain must then correlate the patterns of activity in the different areas or subareas. Philips *et al.* (1984) refer to the role of the cortex in detecting covariation by representing together in the cortex types of event that are related. Because of the hierarchical organization higher-order areas can compare information that was initially analysed in separate regions. The detection of distant relationships is a challenge to cortical mechanisms.

I am grateful to Mr A. G. M. Canavan for helpful comments on the manuscript. The work was supported by M.R.C. grants 971/1/397B and G8122258N.

References

Cohen, S. M. 1972 Electrical stimulation of cortical–caudate pairs during delayed successive visual discrimination in monkeys. *Acta neurobiol. exp.* **32**, 211–233.

Collin, N. G., Cowey, A., Latto, R. & Marzi, C. 1982 The role of frontal eye-fields and superior colliculi in visual search and non-visual search in rhesus monkeys. *Behav. Brain Res.* **4**, 177–193.

Cowey, A. 1968 Discrimination. In *Analysis of behavioral change* (ed. L. Weiskrantz), pp. 189–238. New York: Harper and Row.

Eisenberg, J. F. 1981 *The mammalian radiations*. London: Athlone Press.

French, G. M. 1964 The frontal lobes and association. In *The frontal granular cortex and behavior* (ed. J. M. Warren & K. Akert), pp. 56–72. New York: McGraw-Hill.

Goldman, P. S. 1974 An alternative to developmental plasticity: heterology of CNS structures in infants and adults. In *Plasticity and recovery of function in the central nervous system* (ed. D. G. Stein, J. J. Rosen & N. Butters), pp. 149–174. New York: Academic Press.

Goldman, P. S. & Alexander, G. E. 1977 Maturation of prefrontal cortex in the monkey revealed by local reversible cryogenic depression. *Nature, Lond.* **267**, 613–615.

Goldman, P. S., Rosvold, H. E., Vest, B. & Galkin, T. W. 1971 Analysis of the delayed-alternation defecit produced by dorsolateral prefrontal lesions in the rhesis monkey. *J. comp. Physiol. Psychol.* **73**, 212–220.

Halsband, U. 1982 Higher movement disorders in monkeys. Unpublished D.Phil. Thesis, University of Oxford

Halsband, U. & Passingham, R. E. 1982 The role of premotor and parietal cortex in the direction of action. *Brain Res.* **240**, 368–372.

Harlow, H. F. & Mears, C. 1979 *The human model: primate perspectives*. New York: Wiley.

Harman, P. J. 1947 Quantitative analysis of the brain-isocortex relationship in Mammalia. *Anat. Rec.* **97**, 342.

Honig, W. K. 1978 Studies on working memory in the pigeon. In *Cognitive processes in animal behavior* (ed. S. H. Hulse, H. Fowler & W. K. Honig), pp. 211–248. New York: Academic Press.

Jarvik, M. E. 1953 Discrimination of colored food and food signs by primates. *J. comp. Physiol. Psychol.* **46**, 390–392.

Jones, E. G. & Powell, T. P. S. 1970 An anatomical study of converging pathways within the cerebral cortex of the monkey. *Brain* **93**, 793–820.

King, J. E. & Fobes, J. L. 1982 Complex learning by Primates. In *Primate behavior* (ed. J. L. Fobes & J. E. King), pp. 327–360. New York: Academic Press.

Kojima, S., Kojima, M. & Goldman-Rakic, P. S. 1982 Operant behavioral analysis of memory loss in monkeys with prefrontal lesions. *Brain Res.* **248**, 51–59.

Kruska, D. 1975 Vergleichend-quantitative Untersuchungen an den Gehirnen von Wander- und Laborratten. 1. Volumenvergleich des Gesamthirns und der klassichen Hirnteile. *J. Hirnforsch.* **16**, 469–483.

Latto, R. M. 1982 Visual perception and oculomotor areas in the primate brain. In *Advances in the analysis of visual behaviour* (ed. D. J. Ingle, R. J. W. Mansfield & M. A. Goodale), pp. 671–691. Cambridge, Massachusetts: MIT Press.

Lawler, K. A. 1981 Aspects of spatial vision after brain damage. Unpublished D.Phil. thesis, University of Oxford.

McGonigle, B. O. & Flook, J. 1978 The learning of hand preferences by squirrel monkey. *Psychol. Res.* **40**, 93–98.

Milner, A. D., Foreman, N. P. & Goodale, M. A. 1978 Go-left go-right discrimination performance and distractibility following lesions of prefrontal cortex or superior colliculus in stumptail monkeys. *Neuropsychologia* **16**, 381–390.

Mishkin, M., Spiegler, B. J., Saunders, R. C. & Malamut, B. C. 1982 An animal model of global amnesia. In *Alzheimer's disease* (ed. S. Corkin, K. L. Davis, T. H. Growdon, E. Usdin & R. J. Wurtman), pp. 235–247. New York: Raven.

Pandya, D. N. & Seltzer, B. 1982 Association areas of the cerebral cortex. *Trends Neurosci.* November, 386–390.

Passingham, R. E. 1971 Behavioural changes after lesions of frontal granular cortex in monkeys (*Macaca mulatta*). Unpublished Ph.D. thesis, University of London.

Passingham, R. E. 1975 Delayed matching after selective prefrontal lesions in monkeys (*Macaca mulatta*). *Brain Res.* **92**, 89–102.

Passingham, R. E. 1978 Information about movements in monkeys (*Macaca mulatta*). *Brain Res.* **152**, 313–328.

Passingham, R. E. 1985*a* The memory of monkeys (*Macaca mulatta*) with lesions in prefrontal cortex. *J. comp. Physiol. Psychol.* (In the press.)

Passingham, R. E. 1985*b* Cues for action in monkeys (*Macaca mulatta*) with lesions in premotor cortex. (In preparation.)

Passingham, R. E., Rawlins, J. N. P., Lightfoot, V. & Fearn, S. 1985 The functional organization of frontal cortex in the rat. (In preparation.)

Petrides, M. 1982 Motor conditional associative-learning after selective prefrontal lesions in the monkey. *Behav. Brain Res.* **5**, 407–413.

Philips, C. G., Zeki, S. & Barlow, H. B. 1984 Localization of function in the cerebral cortex. *Brain* **107**, 328–361.

Rawlins, N. 1985 The hippocampus as a temporary memory store. (In preparation.)

Rizzolatti, G., Matelli, M. & Pavesi, G. 1983 Defecits in attention and movement following the removal of postarcuate (area 6) and prearcuate (area 8) cortex in macaque monkeys. *Brain* **106**, 655–673.

Stephan, H., Frahm, H. & Baron, G. 1981 New and revised data on volumes of brain structures in insectivores and primates. *Folia Primatologia* **35**, 1–29.

Thomas, R. K. 1980 Evolution of intelligence: an approach to its assessment. *Brain Behav. Evol.* **17**, 454–472.

Warren, J. M. 1965 Primate learning in comparative perspective. In *Behavior of nonhuman primates* (ed. A. M. Schrier, H. F. Harlow & F. Stollnitz), vol. 1, pp. 249–281. New York: Academic Press.

Wechsler, D. 1945 A standardized memory scale for clinical use. *J. Psychol.* **19**, 87–95.

Weiskrantz, L. 1982 Comparative aspects of studies of amnesia. *Phil. Trans. R. Soc. Lond.* B **298**, 97–109.

Zimmermann, R. R. & Torrey, C. C. 1965 Ontogeny of learning. In *Behavior of nonhuman primates* (ed. A. M. Schrier, H. F. Harlow & F. Stollnitz), vol. 2, pp. 405–447. New York: Academic Press.

Phil. Trans. R. Soc. Lond. B **308**, 113–128 (1985) [113]
Printed in Great Britain

Animal cognition: thinking without language

By H. S. Terrace

Department of Psychology, Columbia University, New York, New York 10027, *U.S.A.*

Recent attempts to teach apes rudimentary grammatical skills have produced negative results. The basic obstacle appears to be at the level of the individual symbol which, for apes, functions only as a demand. Evidence is lacking that apes can use symbols as names, that is, as a means of simply transmitting information. Even though non-human animals lack linguistic competence, much evidence has recently accumulated that a variety of animals can represent particular features of their environment. What then is the non-verbal nature of animal representations? This question will be discussed with reference to the following findings of studies of serial learning by pigeons. While learning to produce a particular sequence of four elements (colours), pigeons also acquire knowledge about the relation between non-adjacent elements and about the ordinal position of a particular element. Learning to produce a particular sequence also facilitates the discrimination of that sequence from other sequences.

1. Introduction

The existence of human thought is self-evident. Whether animals think is more difficult to determine. Unlike human beings, animals cannot speak about their thoughts. The only evidence that animals think is indirect evidence which must be gleaned from their behaviour.

Scientists and laymen approach such evidence with deep-seated and contradictory attitudes. One point of view, usually attributed to Descartes, regards animals as unthinking mechanical beasts. Behaviour is elicited automatically by stimuli that originate in the animal's internal or external environments. However complex or elaborate the animal's behaviour, it can always be reduced to some configuration of reflexes in which thought plays no role. Descartes also argued that an animal, no matter how intelligent, lacked the capacity to learn language, the vehicle of human thought.

Darwin's theory of evolution acknowledges the possibility that animals can think. In comparing man and animals, Darwin argued that it is just as logical to say that the human mind evolved from animal minds as to say that human anatomical and physiological structures evolved from their animal counterparts.

Until recently, there has been little concrete basis for choosing between the contradictory positions of Descartes and Darwin. During the past two decades, however, students of animal behaviour have provided two important answers to the venerable question 'can an animal think?'. For the first time, ample evidence is available of the existence of animal thought (see, for example, Hulse *et al.* 1978; Roitblat *et al.* 1984. There is considerable agreement that apes, considered by many psychologists to be man's most intelligent relatives, are unable to master the basic features of human language (Premack 1979; Savage-Rumbaugh *et al.* 1980*a*; Terrace *et al.* 1979). That state of affairs raises a fascinating question that can be asked of animals in general: 'what is the non-linguistic medium of animal thought?'.

This question poses obvious problems of definition regarding thought and language. In a later section of this paper, I will review some examples of cognitive processes in animals and indicate why any definition of animal cognition must concern itself with representations of previous experiences that an animal can generate to solve some current problem. That is, to argue that an animal thinks is to have evidence that, in the absence of an external stimulus, an animal learns to represent to itself some feature of a prior experience that will serve as a cue for an appropriate choice of response (cf. Hunter 1913; Terrace 1984*a*). For the moment, I will focus on language and try to justify my claim that it is absent in animals. Once we understand what is absent, we will, at the very least, see the importance of developing non-linguistic models of animal cognition.

2. Original goals of ape language projects

Though linguists, philosophers, psycholinguists and psychologists have yet to agree on a rigorous definition of language that encompasses its many complexities, there is general agreement that the most distinctive feature of human languages is the provision they make for creating new meanings by combining arbitrary words into sentences according to arbitrary grammatical rules. In contrast to the fixed character of various forms of animal communication, the meaning of a word is arbitrary. A sentence characteristically expresses a semantic proposition through words and phrases, each bearing a well-defined but nevertheless arbitrary relation to one another (for example, in some languages actions can precede objects; in others, actions follow objects).

Our ability to create and comprehend novel sentences has prompted many linguists (for example, Chomsky 1966) to argue that human grammatical competence is innate and species-specific. It was in this neo-Cartesian Zeitgeist that the various recent ape language projects were started (see, for example, Gardner & Gardner 1969; Premack 1971). Accordingly, it was not surprising that a general goal of those projects was the demonstration of grammatical competence in apes. As we shall later see, that goal took too much for granted regarding an ape's non-grammatical linguistic competence and stimulated the various ape language projects to set goals for themselves that were unrealistically ambitious. It is only recently that researchers in this area have redirected their efforts toward more productive lines of inquiry (for example, Savage-Rumbaugh *et al.* 1983).

The initial results of the various ape-language projects produced exciting evidence of an ape's ability to create sentences. (See Ristau & Robbins 1982 for a thorough summary of the literature on attempts to teach language to apes.) For example, in an early diary report, the Gardners noted that Washoe used her signs 'in strings of two or more...in 29 different two-sign combinations and four different combinations of three signs'. That report prompted Brown to comment 'It was rather as if a seismometer left on the moon had started to tap out "S–O–S"' (Brown 1970, p. 211). Indeed, Brown compared Washoe's sequences of signs to the early sentences of a child, in particular, with respect to their structural meanings (for example, agent–action, agent–object, action–object, and so on).

Other projects reported similar combinations of two or more symbols. Sarah produced strings of plastic chips such as MARY GIVE SARAH APPLE (Premack 1976). Rumbaugh *et al.* (1973) taught a juvenile female chimpanzee named Lana to use an artificial language of 'lexigrams'. Each lexigram was constructed by superimposing an arbitrary geometric configuration on one of six coloured backgrounds. After learning to use individual lexigrams, Lana was taught to produce sequences such as PLEASE MACHINE GIVE M & M. Subsequently, Patterson (1978) reported that Koko, a young female gorilla she taught to use American Sign Language (A.S.L) produced many combinations of two or more signs.

The imitative and non-spontaneous nature of an ape's signing

By 1980, it became apparent that the evidence purporting to show that an ape could create a sentence could be explained without any reference to grammatical competence. Terrace (1979; Terrace *et al.* 1979) analysed approximately 20000 combinations of two or more signs made by a young chimpanzee (Nim) who, like Washoe, had been reared by his human surrogate parents in an environment in which A.S.L. was the major medium of communication. Superficially, many of Nim's combinations appeared to be generated by simple finite-state grammatical rules (for example, MORE + X; transitive verb + me or Nim, etc.). However, a frame-by-frame analysis of videotapes of Nim's signing with his teachers revealed that Nim responded mainly to the urgings of his teacher to sign and that much of what he signed was a full or partial imitation of his teacher's prior utterance. Thus, unlike a child at stage 1 of language acquisition (cf. Brown 1973; Bloom *et al.* 1976, Nim's signing was mostly non-spontaneous and imitative. Analyses of the available films of other signing apes revealed similar patterns of non-spontaneous and imitative discourse (for example, Washoe signing with the Gardners and her other teachers, and Koko signing with Patterson).

The conclusions of project Nim were criticized by other investigators attempting to teach an ape to use sign language on various methodological grounds (for example, Gardner 1981; Patterson 1981). However, none of those investigators have revealed enough of their own procedures to allow one to evaluate the significance of their criticisms of project Nim (Terrace *et al.* 1981; Terrace 1982). Of greater interest is the fact that Terrace's conclusions have yet to be countered with positive evidence. Specifically, no discourse analysis of an ape's signing (as obtained from unedited videotape or film records) has been presented which shows that an ape's utterances are spontaneous and that they are not whole or partial imitations of the teacher's most recently signed utterance.

Rote sequences versus sentences

Different considerations led to a rejection of the view that Sarah's and Lana's sequences were sentences. After analysing a corpus of approximately 14000 of Lana's combinations that were collected by a computer, Thompson & Church (1980) concluded that those combinations could be accounted for almost completely by two non-grammatical processes; conditional discrimination and paired-associate learning. Which of six stock sentences occurred could be predicted by the circumstances under which Lana would try to obtain some incentive. For example, if the object was in view in the machine, the stock sequence would be of the form PLEASE MACHINE GIVE X or PLEASE MACHINE GIVE PIECE OF X. If there was no object in view, the appropriate sequence would be PLEASE PUT INTO MACHINE X. If an experimenter was present, the stock sequence would be of the form PLEASE Y GIVE X. In addition, Lana learned paired associates, each consisting of a particular lexigram and a particular incentive (for example, apple, music, banana, chocolate, and so on). These lexigrams were inserted in the appropriate position (usually the last) of the stock sentence.†

Further evidence of the non-sentential nature of Lana's (and Sarah's) sequences were produced by studies demonstrating that pigeons could be trained to respond in an arbitrary sequence to four simultaneously presented coloured lights (red, green, yellow and blue) whose positions was changed from trial to trial (Straub *et al.* 1979; Terrace 1984*b*). Mention of a pigeon's sequence-learning ability is not to imply that a pigeon could approach a chimpanzee's ability to learn various conditional discriminations which specify which arbitrary sequence is to be emitted in which context. Nor is it meant to imply that a pigeon could master even a single sequence as rapidly as could a chimpanzee. Indeed, there is strong evidence to the contrary (Pate & Rumbaugh 1983). There is also no reason to assume that pigeons and chimpanzees use similar strategies in learning to produce a sequence.

There considerations should not, however, detract from the fact that, in each case, what was learned was a rote sequence. It would be just as erroneous to interpret a rote sequence of pecks to the colours, red, green, yellow and blue (in that order) as a sentence meaning PLEASE MACHINE GIVE GRAIN as it would be to interpret the sequence that a person produces while operating a cash machine as a sentence meaning PLEASE MACHINE GIVE CASH. In sum, a rote sequence, however that sequence might be trained, is not necessarily a sentence.

What do the words of an ape's vocabulary mean?

In a searching review of their own work and that of other projects, Savage-Rumbaugh *et al.* (1980*a*) not only doubted the validity of evidence purporting to show that apes can produce and comprehend sentences but also doubted whether, at the level of individual elements of their vocabularies, the apes studied by any project used those elements as actual words (Lana included).

By questioning the lexical status of an ape's use of signs, of A.S.L., of plastic chips or of lexigrams, Savage-Rumbaugh *et al.* identified a basic problem of interpretation that is common to all of the projects that sought to demonstrate that apes could master simple features of human languages. Indeed a strong case can be made for the hypothesis that the deceptively simple ability to use a symbol as a name required a cognitive advance in the evolution of human intelligence that was at least as significant as the advance(s) that led to grammatical competence.

The development of a child's vocabulary: the behaviourist view

Thanks, in large part, to a preoccupation with the emergence of grammatical competence in children, developmental psycholinguistics have paid relatively little attention to the *process* of lexical acquisition *per se*. It is, of course, true that ample information is available regarding the kinds of words children learn and at what rate they do so (for example, Brown 1956; Clark

† The validity of Thompson & Church's 'stock sentence plus paired associate' hypothesis was questioned by Pate & Rumbaugh 1983 in their analysis of Lana's later productions. Pate & Rumbaugh concluded that, given the variety of the stock sentences and the paired associates that would be needed to account for Lana's well-formed sentences, it would seem reasonable to postulate that they were constructed by a more complex (but unspecified) set of rules. That conclusion seems premature for a variety of reasons. Of the 881 sentences that were analysed only 512 (58.1 %) were well-formed. Further, many of the sequences were quite similar to one another and variation of a particular sequence appeared to be synonymous, for example, JUICE NAME THIS, JUICE NAME THIS IN CUP, JUICE NAME THIS THAT'S IN CUP IN ROOM, etc. Given the constant context, longer sequences do not appear to add any additional information and may have resulted from the experimenter's implicit encouragement to produce long sequences (cf. Terrace 1979).

1973; Nelson 1973). However, it was widely assumed by most psychologists, at least implicitly, that some version of a 'behaviourist' explanation of vocabulary growth would suffice to explain how children learned their initial vocabulary.

At first glance this might seem to be a reasonable state of affairs since there is general agreement that, unlike sentences, words are learned individually. Accordingly, why not invoke principles of associative learning to account for vocabularly acquisition? On this view, a child associates the speech of a parent with primary reinforcers such as physical contact, food, or the removal of distressful stimulation. As a consequence, the parent's vocalizations become reinforcing.

At the same time, the child's vocalizations are presumed to be reinforced directly, either by the parent providing a primary reinforcer, by the parent's attention or by the parent's vocalization. Initially, virtually any instance of an infant's babbling is reinforced. As the infant becomes older, the parent shapes her vocabulary to approximate adult sounds. In addition, those infant's vocalizations that resemble parent's speech are self-reinforcing. Gradually, the frequency of the infant's vocalizations that resemble sounds uttered by a parent increases while the frequency of those sounds which differ from the sounds uttered by the parent decreases (for example, Mowrer 1954; Winitz 1969).

Naming versus paired-associate learning

At best, the behaviorist view of vocabulary acquisition is an explanation of paired-associate learning: learning to use an arbitrary symbol as a means towards the end of obtaining some reinforcer in the presence of a particular stimulus. What is missing from the behaviourist view is the speaker's intention in using a word. Saying something and meaning what you say are obviously two different kinds of response. In most human discourse, a speaker who utters a name expects the listener to interpret the speaker's utterance as the referent for a jointly perceived object. It should therefore come as no surprise that the function of much of a child's initial vocabulary of names is to inform another person, usually a parent, that the child has noticed something (Halliday 1975; MacNamara 1982). In many instances, the child refers to the object in question spontaneously, with obvious delight, and shows no interest in obtaining the object. The child appears to not only enjoy sharing information with her parent but to also derive intrinsic pleasure from the sheer act of naming. As I will argue later, these aspects of uttering a name have not been observed in apes and there is reason to doubt whether the most intensive training programme imaginable could produce an ape who would approximate a child's natural ability to refer to objects as an end in itself.†

How children learn to use language

An obvious truism about language learning is that it draws upon certain kinds of non-linguistic knowledge. For example, before learning to speak, an infant acquires a repertoire of instrumental behavior that allows her to manipulate or approach various objects. An infant also learns how to engage in various kinds of social interaction with her parents, for example,

† Other deficiencies of the behaviourist view of word acquisition can be found in studies showing that there is little evidence that the sounds an infant emits are truly imitative of the parent's sounds (see, for example, Winitz & Irwin 1958) and that, in many instances, a child's initial utterances function as names rather than as requests (for example, Nelson 1973). Discussions of these and other observations that cannot be accommodated by a behaviourist account of language development can be found in various recent accounts of the growth of a child's vocabulary (for example, Bloom & Lahey 1978; Nelson 1973).

being able to look where the parent is looking or pointing.† Eventually, the child learns to point at things that she would like her parent to notice. In short, the infant first masters a social and conceptual world onto which she can later map various kinds of linguistic expression.

The rapidly expanding literature on the pre-linguistic development of the child makes clear that, for whatever reason, or reasons, naming emerges from the highly structed interactions between an infant and her parents. Especially relevant are interactions in which the parent is able to direct the infants's attention to particular objects. For example, at the age of roughly four months, a parent can direct an infant's attention to an object simply by looking at it. Subsequently, the parent can accomplish the same end by pointing to an object. Often the parent will comment about the object while pointing to it or moving it towards the infant. By placing stress on the spoken name of the object to which the parent seeks to direct the infant's attention, the infant comes to discover that a stressed vocalization is a signal that there is 'something to look at'. Likewise, highly ritualized games whereby an object is made to disappear and later reappear (typically, with distinctive vocal accompaniments) also facilitates a parent's control over an infant's attention. As the infant gets older, her contribution to these interactions increases. At first she may only point to an object in response to the parent's pointing or vocalizing. Subsequently, the child may utter non-standard vocalizations while looking or pointing at the object presented by the parent. Eventually, the child learns to repeat the object's name as provided by the parent, while the child and the parent jointly attend to that object (see Bruner (1983) for summaries of such studies).

During the course of a long series of object-oriented interactions with her parents, an infant not only learns to direct her attention to objects that are presented by her parent but she also learns that her response to such objects, whether pointing, babbling or saying the actual name of the object, is recognized by the parent as a sign that she has noticed the object. In short, the infant learns that her response to an object has much in common with her parent's response to the same object. In that sense, the child learns the conventions of reference, first non-verbally and subsequently at a verbal level.

Can referring be taught?

In a provocative discussion of how children learn to name objects, MacNamara (1982) concludes that referring to an object (the act of communicating that one's attention is directed to a particular object) is not learned. Instead he regards referring as a 'primitive of cognitive psychology' (MacNamara 1982, p. 190). What is learned is reference: the conventions of using symbols and words that do the work of referring.

Verification of MacNamara's view of learning names awaits much further research, his painstaking marshalling of empirical and logical arguments notwithstanding. It is of interest, however, to consider the extent to which learning theory can account for a child's ability first, to understand that her parent is referring to a particular object and, subsequently, to master pre-verbal techniques for directing her parent's attention to a particular object. As commonplace as such skills may seem, it is not obvious how one can teach them. To argue that referential behaviour is shaped begs the question of what rudimentary forms of referential behaviour can be used as a point of departure for the shaping process. To acknowledge that such a rudimentary

† It is clearly more generally accurate to invoke shared attention than shared looking in describing pre-linguistic origins of reference. For example, 'looking' would be inappropriate as a description of how a blind child responds to particular objects. Nevertheless, blind children develop a sense of the focus of their parents' attention and readily learn to refer to the act of engaging someone else's attention with verbalisms such as 'look' (Landau & Gleitman 1983).

form exists, is to agree with MacNamara that the act of referring is a given. At best, principles of learning might be invoked to characterize how a parent adds to the variety and complexity of situations in which referring occurs.

The function of symbols for chimpanzees and children

The hypothesis that the act of referring is a given and that it is also a necessary precursor of naming provides an important basis for comparing symbol use by children and chimpanzees. Like children, chimpanzees appear to show evidence of object-recognition soon after birth. It is also quite easy to direct their attention to a particular object by looking at it, by pointing to it or by moving it into the chimpanzee's line of sight. Though their reactions to objects have not been subjected to systematic study, informal observations suggest that their main reaction is acquisitive (Terrace 1979). When confronted with an object, familiar or otherwise, an infant ape will make soft reflexive hooting noises and either reach for the object or try to approach it. Typically, the object is explored orally and manually. However, beyond such explorations there is no evidence that suggests that an infant ape is interested in communicating, to another ape or to its human surrogate parent, the fact that it has noticed an object, as an end in itself. To be sure, chimpanzees will communicate with one another about food locations (cf. Menzel 1979) or about objects of prey (cf. Telecki 1973). It is, however, important to recognize that such communication is in the service of some concrete end and is not intended simply to inform a companion that something has been noticed.

The absence of natural referential skills that are independent of concrete ends makes all the more remarkable the kinds of symbol use that an ape can master. For example, recent studies have shown that chimpanzees are capable of learning symbolic concepts such as generic terms that apply to symbols for particular foods and tools (Savage-Rumbaugh *et al.* 1980*b*). They have also shown some rudimentary intentional communication in highly structured play situations (Savage-Rumbaugh *et al.* 1983).

Naming as a precursor of syntax

It is beyond the scope of this paper to speculate when in the evolution of human intelligence infants were able to relax their acquisitive reactions to objects and simply indicate to a parent that they noticed an object. Whatever the origin of that kind of reaction, it clearly exerted a significant influence on the evolution of language. Foremost, it provided a psychological basis for activities between an infant and her parent for engaging in activities based upon their joint perception of an object. As we have seen, such activities are important precursors of reference to objects and events with names.

If a child hadn't developed the ability to use a name to register what she saw and if the sole function of her speech was to demand things, it is hard to see why she would combine words according to a grammatical rule. Since a single word should suffice as a demand or as a warning of some danger, the child would have no need to learn to speak syntactically. Obviously, the same argument applies to apes and indicates why it was premature to have expected that an ape might master even the most primitive grammatical rules.

In theory, one could, of course, argue that a highly structured system of demands might require syntactic rules, for example, a request for the red plum from the far tree, as opposed to the green apple under the near bush, and so on. Such a state of affairs is implausible for a variety of reasons. To the extent that such specific desires occur in the natural world, they could be dealt with by eye–gaze, pointing, facial expressions, some combination thereof, or by a process of elimination of alternative incentives. Thus, it is not clear what natural function a hypothetical demand system of such complexity might serve. Further, any attempt to teach such skills in a laboratory environment would seem to tax the ability of any known primate other than man.

A different state of affairs exists in those situations in which there is a desire simply to transmit information about a relationship between one object or action and another, or about some attribute of an object or about past or future events. In these instances, a single word would not suffice. Hence, the functional value of syntax.

Representations as evidence of animal thinking

For reasons far more elementary than those advanced by the contemporary neo-Cartesian school of linguistics, we have seen that Descartes was correct in denying that animal lack the capacity to learn a human language. However, Descartes' contention that animals cannot think was based as much on their inability to master a language as it was on his view that their behaviour consisted of nothing more than a mechanical system of reflexes. That view in particular became the creed of 20th century behaviourists (for example, Pavlov 1927; Guthrie 1952; Hull 1943; Skinner 1938). It is, of course, true that behaviourist developed models of reflexes (and their combination) that were more elaborate than the model suggested by Descartes, particularly in the case of learned behaviour. It should also be obvious that their models assumed a more realistic view of the nervous system than was available to Descartes. However, like Descartes, modern behaviourists saw no need to appeal to cognitive structures that intervened between a stimulus and a response so long as their models of conditioned behaviour could predict reliably the occurrence of a particular response.

An important tension in the modern study of behaviour is one that resulted from a tug of war between behaviourists and cognitively oriented psychologists as to validity of instances of animal behaviour that were purported to be exceptions to reflex models of behaviour. The significance of such exceptions was recognized more than 70 years ago by Hunter, an early behaviourist, who observed that:

> '...if comparative psychology is to postulate a representative fact,...it is necessary that the stimulus represented be absent at the moment of response. If it is not absent, the reaction may be stated in sensory-motor terms' (Hunter 1913, p. 21).

By stipulating that '...the stimulus represented be absent at the moment of the response', Hunter argued that the only cue available to the organism was one that it generated as some representation of the absent stimulus. That representation functions just as an exteroceptive stimulus might in evoking appropriate behaviour.†

To grasp fully Hunter's view of an animal's ability to use a self-generated representation of some feature of its environment, it is important to ask why he did not see any need to argue for representation in those instances in which a stimulus reliably precedes a response. Why not,

† In the various types of delayed response apparatus that Hunter devised to study representation in animals, he took pains to rule out particular postures or orientations of the organism as mediators between the stimulus and the response recorded by the experimenter. Hunter correctly rejected, as representations, kinesthetic feedback from such mediators because they could be construed as members of a covert chain of stimuli and responses. Overall, Hunter found very little evidence that animals could represent features of their environment in ways that could not be explained in S–R terms.

for example, appeal to representations of a conditioned stimulus or of an unconditioned stimulus in a typical conditioning experiment?† Hunter's answer was the logic of parsimony. Like Skinner and other behaviourists, Hunter noted that our ability to predict or to explain behaviour is not enhanced by appealing to a representation of a stimulus if that stimulus is available when the organism responds.

This is not to say that, somehow, an animal doesn't store memories of its experience. Quite obviously it must if it is able to react to a stimulus at time 2 in a manner similar to its reaction to that stimulus at time 1. It is, however, necessary to distinguish between an organism's ability to generate, or at least to maintain, a representation of some previously experienced stimulus that is present when the response in question occurs and its ability to respond when that stimulus is absent.

More so than any other type of study, recent experiments on animal memory provide compelling evidence that animals form representations. In each case the subject is required to recall certain features of one or more events and, in the absence of those events, to use that information as a basis for performing some response.

Learned sequences of responses: the traditional view

Of particular interest are experiments involving integrated sequences or responses. Though behaviour typically occurs in integrated sequences, learning theorists have concerned themselves mainly with principles of conditioning as they bear on individual repetitive responses such as bar-presses and key-pecks and how such responses are influenced by the presence of a particular cue or the value of a particular schedule of reinforcement. Integrated sequences of response are regarded as 'chains' of discrete responses (cf. Skinner 1938; Hull 1943; Guthrie 1952; Spence 1956; Logan 1960). This model of behavior, which derives from Sherrington's (1906) formulation of chains of reflexes, assumes that the stimulus consequences of one response function as a cue for the next response. On this view, an organism who learns a sequence of responses has simply learned to respond appropriately to a series of successively presented stimuli, and nothing more. For example, a rat that learns to run through a maze need not have any knowledge of the plan of the maze. At first choice point, it makes a response appropriate to S_1; the response, R_1, is followed by the appearance of S_2, and so on.

The literature on chaining in animals consists almost entirely of experiments in which the subject encounters, one at a time, the stimuli that occasion successive responses and the manipulanda for responding instrumentally to those stimuli. In a maze, for example, a rat progresses through a sequence of choice points. Each choice point can function as a unique discriminative stimulus for the next response. In operant conditioning chambers, the acts of a rat pressing a bar or of a pigeon pecking a key have been conceptualized as a sequence of responses (approach, postural shifts, turning the head, etc.), each of which produces a new discriminative stimulus (Skinner 1938; Skinner 1953; Keller & Schoenfeld 1950). Common to each of these examples is an organizational plan whereby a particular response results in the elimination of one discriminative stimulus and the presentation of a new one. In turn, the presentation of each new discriminative stimulus provides an opportunity to make a response that could not have occurred earlier.

† It is immaterial, for this argument, whether the representation is of the CS, the US, the UR, or of some S–S or S–R connection. The form of the representation is less critical than the fact that some type of representation exists.

3. Response sequences that traditional chaining theory cannot explain

The radial maze

Olton's experiments on a rat's behaviour in a radial maze provide an important departure from traditional maze studies. Consider a radial maze in which eight runways radiate from a common start point. Olton & Samuelson (1976) and Olton (1978; 1979) demonstrated that rats can remember which of the runways they entered while searching for food. In the basic paradigm, each runway was identical and was baited with equal amounts of food that could not be seen from the entrance to the runway. The variable of interest was how many runways were entered before all the food was consumed. After a few days in the maze, Olton's rats, on the average, re-entered less than one ally per trial.

To appreciate the performance of Olton's rats, let us identify each of the arms of a radial maze by a number, from 1 to 8. On one trial, the rat's sequence of arm entries might be 3-7-2-1-5-4-6-8; on another, it might be 6-1-5-7-3-8-4-2; and so on. Thus, it is impossible to predict the rat's sequence on one trial from the sequence followed on previous trails. This poses an interesting question. How does the rat remember which arms it has already visited?

One possibility is that, as the rat enters or leaves an arm, it lays down some kind of trail that it can detect when it returns to the entrance of that arm. This explanation was ruled out by the results of a number of subsequent experiments. In one instance a highly odorous substance was used to mask presumed olfactory cues. In another, the olfactory nerve was cut. Neither procedure impaired the rat's efficient performance. In another clever variation of the basic training procedure, the rat was confined on the centre platform after it made a few choices. The arms of the maze were rotated some arbitrary distance, say, 90°. The rat then visited arms that now occupied the *position* of previously unvisited arms, even though, because of the rotation, it had, in fact visited some of them before. Taken together, these observations show that the rat's choices cannot be attributed to any particular external cue. In that sense, they are not conditioned responses to some feature of the environment.

A rat's ability to create and use a map of food locations is clear evidence that, while foraging, it is not simply making a series of mindless responses, each triggered by some environmental cue. But however remarkable this ability may seem, it sheds no light on an animal's ability to learn *particular* sequences. Young children, for example, find it easy to learn all kinds of arbitrary sequences by rote, such as nursery rhymes or telephone numbers.

Serial learning in the pigeon

Those kind of sequences, which also resemble the rote sequences chimpanzees were trained to perform to obtain some reward (for example, MARY GIVE SARAH FOOD), were studies in a series of experiments in which pigeons served as subjects. The procedure used to train these sequences differed significantly from conventional chaining paradigms. Such paradigms do *not* require the organism to memorize the sequence that defines a particular chain. For example, in learning to run through a typical maze, a rat has only to learn what to do at various choice points. Since each choice point is discriminable, the rat's task can be characterized as learning to solve a set of discrimination problems in which the discriminative stimuli are encountered successively.

In contrast to the traditional successive-chaining paradigm, in which discriminative stimuli and the opportunity to make a particular response are encountered one at a time, a simultaneous chaining paradigm presents, at the same time, all of the stimuli and all of the manipulanda for each response. Another essential feature of the simultaneous-chaining paradigm is that it provides no differential step-by-step feedback following each response. In contrast, a conventional successive-chaining paradigm ensures that each correct response produces feedback that typically results in the automatic replacement of the current discriminative stimulus by the next one.

In the first studies to employ a simultaneous chaining paradigm (Terrace *et al.* 1977; Straub *et al.* 1979), pigeons were trained to learn a 'list' of coloured lights. Each trial presented an array of four colours (A, B, C, D) each randomly positioned on different response keys. To obtain food reinforcement, the pigeon had to respond to each array by pecking in the sequence A-B-C-D regardless of how those colours were positioned on the response keys. For example, on one trial the left–right arrangement of the colours might be B, C, A, D; on the next trial it might be D, B, C, A. In each case access to food was provided if and only if the pigeon pecked the keys in the order A-B-C-D. If the subject made an error the array was turned off and a new array was presented during the next trial. With the exception of the response to the last colour (which was followed by reinforcement), no differential feedback was provided following correct responses.

Pigeons learned to perform the A-B-C-D sequence on the arrays on which they were trained at levels of accuracy that exceeded 70%. No decrement in performance was observed even when the colours A, B, C, and D were configured in novel arrays (cf. figure 1). This showed that the pigeon had not simply mastered a set of rotely learned response sequences to the arrays used during training. Thus, the only basis the pigeon had for choosing a particular colour is its representation of what colour it just pecked and what colour should next be pecked.

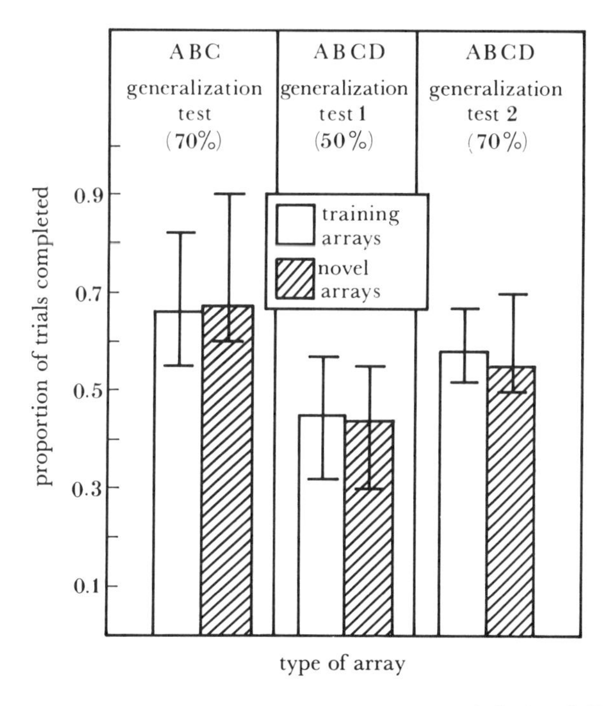

FIGURE 1. Average proportion of training and novel arrays computed during A-B-C generalization test at 70% accuracy level (left-hand panel) and during the A-B-C-D generalization tests at 50% (middle panel) and 70% accuracy levels (right-hand panel). See Straub & Terrace (1981) for further details.

A pigeon's ability to form a representation of the sequence was also demonstrated by its accurate performance on 'sub-sets' of the original sequence (for example, B and D, A and D, C and D, and so on). On arrays presenting B and D (in which the required sequence was B-D), the pigeon had neither the advantage of the normal starting colour nor, having pecked B, the advantage of an adjacent element. Just the same, accuracy of performance was as great on arrays requiring the sequence B-D as it was on arrays requiring the sequence A-B.

Another study showed that the ordinal position of the middle element *per se* controlled performance on three-element sequences of coloured elements (Terrace 1984*b*). Following acquisition of the sequence A-B-C, training commenced on one of three new 'lists' of elements. All of the lists consisted of one old element, B, the middle colour of the original list, and two new elements, X and Y, a white vertical line and a white diamond (on black backgrounds), respectively. For one group, B retained its original position in the new sequence: X-B-Y. For both of the other groups, the position of B was shifted (with respect to A-B-C training). One group was trained on the sequence X-Y-B; the other on the sequence B-X-Y.

As shown in figure 2, the subjects trained on the sequence in which B retained its original position (X-B-Y) learned that sequence more rapidly than the subjects trained on the sequences in which the position of B was changed (X-Y-B and B-X-Y). Conversely, it took longer to acquire the sequences in which the ordinal position of B was changed (X-Y-B and B-X-Y) than it did to acquire the original A-B-C sequence. These results show that in addition to mastering a three-element sequence of colours, pigeons also acquired knowledge about the ordinal position of the second element. That knowledge adds to the complexity of the representation that would be needed to account for performance on simultaneous chaining.

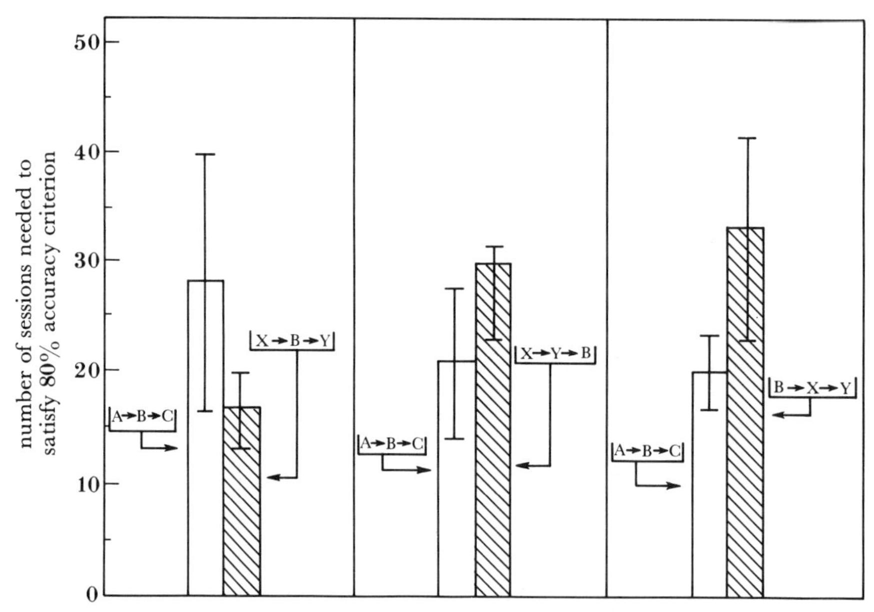

FIGURE 2. Number of sessions needed to acquire X-B-Y, X-Y-B and B-X-Y sequences following training on the simultaneous chain A-B-C.

Yet another consequence of learning to produces a sequence is positive transfer to sequence recognition. In a study recently completed in my laboratory, sequence recognition was trained by a 'yes–no' paradigm. On half of the trials the sequence A-B-C was presented on a single key. On the remaining trials sequences of the elements A, B, and C were presented in other orders. Following each type of sequence, 'test' stimuli were presented on keys to the left and the right of the key on which the sequence was shown. Food reward was provided for pecks to the left hand key following A-B-C sequences and to the right key following non A-B-C sequences (see Dopkins *et al.* (1983); Weisman *et al.* (1980), for similar procedures for training sequence discriminations).

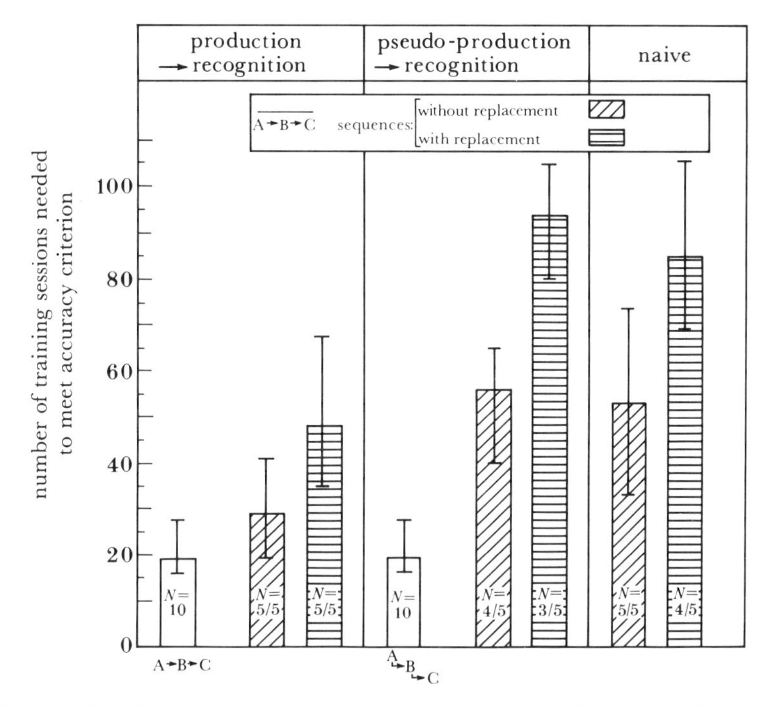

FIGURE 3. Number of sessions to acquire discrimination between the elements A, B and C as presented in A-B-C sequences (left-hand panel), the pseudo-production of A-B-C sequences (middle panel) and in the case of naive subjects (right-hand panel).

The acquisition of sequence discrimination was studied under three conditions. One group of pigeons first learned to produce an A-B-C sequence under the simultaneous chaining procedure described earlier. A second group was given the same amount of training on a successive training procedure in which the elements A, B and C were presented one at a time. (A peck to A produced B and turned off A, a peak to B produced C and turned off B, and a peck to C turned off C and produced food reward.) The third group had no experience with the elements A, B, and C before sequence discrimination training. As shown in figure 3, the first group learned to discriminate the A-B-C from the non A-B-C sequences much more rapidly than the subjects of the other two groups. Thus, the representation that a pigeon acquires when learning to *produce* a sequence facilitates learning to *discriminate* the sequence from other sequences containing the same elements.

4. Difference between human and animal representations

Though our knowledge of animal representations is embarrassingly meagre, we can be fairly confident that animal representations differ from those generated by human beings in two important respects. Most studies of human memory use verbal stimuli. Even when non-verbal stimuli are used, memory may be facilitated by verbal mnemonics and control processes. In the absence of such mnemonics and control processes, it seems foolhardy to assume that animals rehearse stimuli verbally or that there is much overlap between animal and human encoding processes. It also seems clear that cognitive processes in animals may be more limited by biological constraints than their human counterparts.

In the radial maze, for example, an important basis of the rat's ability to avoid previously visited alleys is an unlearned 'win–shift' strategy that it follows when searching for food. While a win-shift strategy is not sufficient to explain the highly efficient performance of Olton's rats, it appears to be a necessary condition. This becomes evident when comparing the radial-maze performance of pigeons, who are 'win–stay' organisms, with that of rats. Pigeons appear to be considerably less efficient than rats in avoiding previously visited alleys (Bond *et al.* 1981). Given the pigeons's ability to home, it seems more plausible to attribute their poor performance in the radial maze to it's 'win–stay' tendency than to a poorer ability to represent spatial locations.

Putting aside the contribution of a win–shift strategy, it is unclear that the rat's ability to perform efficiently a radial maze has very much in common with such superficially similar human abilities as remembering elements of arbitrary lists, for example, which of a group of people have yet to be called, which errands have yet to be performed, and so on. At present, we have no basis for assuming that a rat's ability to keep track of alleys that it has visited could generalize to tasks that require other responses (for example, bar-pressing) or to the many kinds of arbitrary non-spatial tasks that language makes possible in the case of humans.

Virtually all of the examples of representation in animals described earlier warrant similar caution when it comes to extrapolating to human cognitive processes. For example, both Herrnstein and Lea have noted that the processes used by pigeons to form concepts may differ considerably from those used by human subjects (Herrnstein & deVilliers 1980; Lea & Harrison 1978; Morgan *et al.* 1976. In the case of a pigeon's ability to represent a group of elements in performing a serial learning task, it is unlikely that their representations of these elements has much in common with human representations of serially ordered elements. Both involve representation and both involve sequences. There is, however, good reason to assume that, unlike the pigeon, human subjects encode each element of the sequence verbally.

These, and other problems suggested by recent demonstrations of animal cognition, leaves us with a baffling but fundamental question. Now that there are strong grounds to dispute Descartes' contention that animals lack the ability to think, it is appropriate to determine just how an animal does think. In particular, how does an animal think without language? Learning the answer to that question will provide an important biological benchmark against which to assess the evolution of human thought.

Preparation of this manuscript was supported in part by grants from the U.S. National Science Foundation and the Mrs Cheever Porter Foundation and by fellowships from All Souls' College, University of Oxford, U.K., and the Fulbright Commission.

References

Bloom, L. & Lahey, M. 1978 *Language development and language disorders*. New York: John Wiley & Sons.

Bloom, L. M., Rocissano, L. & Hood, L. 1976 Adult-child discourse: developmental interaction between information processing and linguistic knowledge. *Cogn. Psychol.* **8**, 521–522.

Bond, A. B., Cook, R. G. & Lamb, M. R. 1981 Spatial memory and the performance of rats and pigeons in the radial-arm maze. *Anim. Learn. Behav.* **9**, 575–580.
Brown, R. 1956 The original word game. In *A study of thinking* (ed. J. Bruner, J. Goodnow & G. Austin.) Appendix. New York: John Wiley & Sons.
Brown, R. 1970 The first sentences of child and chimpanzee. In *Selected psycholinguistic papers* (ed. R. Brown), pp. 808–822. New York: Macmillan.
Brown, R. 1973 *A first language.* Cambridge, Mass: Harvard University Press.
Bruner, J. S. 1983 *Child's talk.* New York: Norton.
Dopkins, S. C., Scopatz, R. A., Roitblat, H. L. & Bever, T. G. 1983 Encoding and decision processes in the discrimination of 3-item sequences. Proceedings of the 54th annual meeting of the Eastern Psychological Association, Philadelphia, Pa.
Chomsky, N. 1966 *Cartesian linguistics: a chapter in the history of rationalist thought.* New York: Harper & Row.
Clark, E. 1973 Non-linguistic strategies and the acquisition of word meanings. *Cognition* **2**, 161–182.
Gardner, B. T. 1981 Project Nim: Who taught whom? *Cont. Psych.* **26**, 425–426.
Gardner, R. A. & Gardner, B. T. 1969 Teaching sign language to a chimpanzee. *Science, Wash.* **165**, 664–672.
Guthrie, E. R. 1952 *The psychology of learning.* New York: Harper & Row.
Halliday, M. A. K. 1975 *Learning how to mean – explorations in the development of language.* London: Edward Arnold.
Herrnstein, R. J. & deVilliers, P. A. 1980 Fish as a natural category for people and pigeons. In *The psychology of learning and motivation* (ed. G. H. Bower), vol. 14. New York: Academic Press.
Hull, C. L. 1943 *Principles of behavior.* New York: Appleton-Century-Crofts.
Hulse, S. H., Fowler, H. & Honig, W. K. 1978 (ed.) *Cognitive processes in animal behavior.* Hillsdale, N.J.: Lawrence Erlbaum Associates.
Hunter, W. S. 1913 The delayed reaction in animals and children. *Behav. Monogr.* **2**, no. 1 (serial no. 6).
Keller, F. S. & Schoenfeld, W. N. 1950 *Principles of psychology.* New York: Appleton-Century-Crofts.
Landau, B. & Gleitman, L. R. 1983 The language of perception in blind children. University of Pennsylvania, unpublished.
Lea, S. E. G. & Harrison, S. N. 1978 Discrimination of polymorphous stimulus sets by pigeons. *Q. Jl exp. Psychol.* **30**, 521–537.
Logan, F. A. 1960 *Incentive.* New Haven, CT: Yale University Press.
MacNamara, J. 1982 *Names for things.* Cambridge, Massachusetts: Bradford, M.I.T. Press.
Menzel, E. W. 1979 Communication of object-locations in a group of young chimpanzees. In *The great apes* (ed. D. A. Hamburg & E. R. McGowan), pp. 359–371. Menlo Park, California: Benjamin Cummings.
Morgan, M. J., Fitch, M. D., Holman, J. G. & Lea, S. E. G. 1976 Pigeons learn the concept of an 'A'. *Perception* **5**, 57–66.
Mowrer, O. M. 1954 The psychologist looks at language. *Am. Psychol.* **9**, 660–694.
Nelson, K. 1973 Structure and strategy in learning to talk. *Mon. Soc. Res. Child Dev.* **38** (1 ser. no. 149).
Olton, D. S. 1979 Mazes, maps and memory. *Am. Psychol.* **34**, 588–96.
Olton, D. S. 1978 Characteristics of spatial memory. In *Cognitive processes in animal behavior* (ed. S. H. Hulse, H. Fowler & W. K. Honig). Hillsdale, N.J.: Lawrence Erlbaum Associates.
Olton, D. S. & Samuelson, R. J. 1976 Remembrance of places past: spatial memory in rats. *J. exp. Psychol.: Anim. Behav. Process.* **2**, 97–116.
Pate, J. L. & Rumbaugh, D. M. 1983 The language-like behavior of Lana: is it merely discrimination and paired-associate learning? *Anim. Learn. Behav.* **11**, 134–138.
Patterson, F. G. 1978 The gestures of a gorilla: Language acquistion in another pongid. *Brain Lang.* **5**, 72–97.
Patterson, F. G. 1981 Ape language. *Science, Wash.* **211**, 86–87.
Pavlov, I. P. 1917 *Conditioned reflexes.* London: Oxford University Press.
Premack, D. 1971 On the assessment of language competence in the chimpanzee. In *Behavior of nonhuman primates* (ed. A. M. Schrier & F. Stollnitz), vol. 4, pp. 186–288. New York: Academic Press.
Premack, D. 1972 Teaching language to an age. *Scient. Am.* **227**, 92–99.
Premack, D. 1976 *Intelligence in ape and Man.* Hillsdale, New Jersey: Erlbaum.
Premack, D. 1979 Species of intelligence: debate between Premack and Chomsky. *The Sciences* **19**, 6–23.
Ristau, C. A. & Robbins, D. 1982 Language in the great apes: a critical review. *Adv. Study Behav.* **R12**, 141–255.
Roitblat, H. L., Bever, G. T. & Terrace, H. S. 1984 *Animal cognition.* Hillsdale, N.J.: Lawrence Erlbaum Associates.
Rumbaugh, D. M., Gill, T. V. & von Glasersfeld, E. C. 1973 Reading and sentence completion by a chimpanzee (*Pan*). *Science, Wash.* **182**, 731–733.
Savage-Rumbaugh, E. S., Rumbaugh, D. M. & Boysen, S. 1980*a* Do apes use language? *Am. Sci.*, **68**, 49–61.
Savage-Rumbaugh, E. S., Rumbaugh, D. M., Smith, S. T. & Lawson, J. 1980*b* Reference–The linguistic essential. *Science, Wash.* **210**, 922–925.
Savage-Rumbaugh, E. S., Pate, J. L., Lawson, J., Smith, T. & Rosenbaum, S. 1983 Can a chimpanzee make a statement? *J. exp. Psych. Gen.* **112**, 457–492.
Sherrington, C. S. 1906 *Integrative action of the nervous system.* Cambridge: Cambridge University Press.
Skinner, B. F. 1938 *The behavior of organisms.* New York: Appleton-Century-Crofts.

Spence, K. W. 1956 *Behavior theory and conditioning*. New Haven: Yale University Press.

Straub, R. O., Seidenberg, M. S., Terrace, H. S. & Bever, T. G. 1979 Serial learning in the pigeon. *J. exp. Analys. Behav.* **32**, 137–148.

Straub, R. O. & Terrace, H. S. 1981 Generalization of serial learning in the pigeon. *Anim. Learn. Behav.* **9**, 454–468.

Telecki, G. 1973 *The predatory behavior of wild chimpanzees*. Lewisburg, Pa: Buckness University Press.

Terrace, H. S. 1979 *Nim*. New York: Knopf.

Terrace, H. S. 1982 Evidence for sign language in apes: what the ape signed or how well was the ape loved? *Cont. Psychol.* **27**, 67–68.

Terrace, H. S. 1984*a* Animal cognition. In *Animal cognition* (ed. H. L. Roitblat, T. G. Bever & H. S. Terrace). Hillsdale, N.J.: Lawrence Erlbaum Associates.

Terrace, H. S. 1984*b* Simultaneous chaining: the problem it poses for traditional chaining theory. In *Quantitative analyses of behavior* (ed. R. J. Herrnstein & A. Wagner). Cambridge, Mass.: Ballinger Publishing Co.

Terrace, H. S., Petitto, L. A., Sanders, R. J. & Bever, T. G. 1979 Can an ape create a sentence? *Science, Wash.* **200**, 891–902.

Terrace, H. S., Petitto, L. A., Sanders, R. J. & Bever, T. G. 1981 Reply to Patterson. *Science, Wash.* **211**, 87–88.

Terrace, H. S., Straub, R. O., Bever, T. G. & Seidenberg, M. S. 1977 Representation of a sequence by a pigeon. *Bull. Psychon. Soc.* **10**, 269.

Thompson, C. R. & Church, R. M. 1980 An explanation of the language of a chimpanzee. *Science, Wash.* **208**, 313–314.

Weisman, R. G., Wasserman, E. A., Dodd, P. W. & Larew, M. B. 1980 Representation and retention of two-event sequences in pigeons. *J. exp. Psychol. Anim. Behav. Process.* **6**, 312–325.

Winitz, H. 1969 *Articulatory acquisition and behavior*. New York: Appleton-Century-Crofts.

Winitz, H. & Irwin, O. C. 1958 Syllabic and phonetic structure of infant's early works. *J. Speech Hearing Res.* **1**, 260–256.

Phil. Trans. R. Soc. Lond. B **308**, 129–144 (1985) [129]
Printed in Great Britain

Riddles of natural categorization

By R. J. Herrnstein
Department of Psychology, Harvard University, 33 Kirkland Street, Cambridge, Massachusetts 02138, U.S.A.

Pigeons and other animals can categorize photographs or drawings as complex as those encountered in ordinary human experience. The fundamental riddle posed by natural categorization is how organisms devoid of language, and presumably also of the associated higher cognitive capacities, can rapidly extract abstract invariances from some (but not all) stimulus classes containing instances so variable that we cannot physically describe either the class rule or the instances, let alone account for the underlying capacity.

In contrast, with other contingencies of reinforcement, pigeons will not extract abstract rules of categorization; they will instead learn to identify visual stimuli down to small details, and they will retain much of what they learned for a year and more. How animals can shift between abstraction and photographic retention, and whether or not the two modes can be unified under a single theory are questions that help define the boundaries of knowledge about animal intelligence.

When it comes to categorization, nature clearly has a secret. Animals that are celebrated more for their lack of intelligence than the reverse can sort exemplars of such variety that they out-perform the most ambitious computer simulations (or even the most ambitious theories of a simulation). How can animals with such remarkable powers of classification still seem stupid in some sense? That is the main riddle posed by a growing collection of laboratory experiments on visual categorization by pigeons and other animals. An older hypothesis was the animals classify objects in the natural environment by fixing on some specific, single feature, as the tick supposedly does by responding to butyric acid emanating from the flesh of mammals or the male stickleback fish to the red underside of a competing male. The new evidence leaves no doubt that simple stimuli are not a general solution to the riddle of natural categorization. The general solution to the riddle must deal with the enormous physical variety within natural categories and thus may well bear on wider issues relating to natural and artificial intelligence and to the distinctive nature of biological information processing systems. But, at this point, we know more about the dimensions of the riddle than about its solution. This article samples from the findings out of which the riddle arises and then draws such tentative conclusions as are suggested by them.

1. Acquiring and extending a visual category

In one experiment (Herrnstein 1979) pigeons working in a standard chamber were first thoroughly accustomed to pecking at a switch for brief access to feed, then with no additional special training, they were shown 80 photographic slides (35 mm) of natural scenes projected on a small screen next to the response switch. The slides appeared one at a time for varying durations averaging 30 s. Half of the slides contained trees and half did not, but otherwise the slides looked comparable. The tree slides included full views of single and multiple trees, but also various obscure, distant, inconspicuous, and achromatic silhouetted instances. An attempt was made to sample natural settings in New England.

An experimental session consisted of a single complete rotation of a tray containing the 80 slides. In the presence of a tree slide, pecking was intermittently reinforced with brief access to food. Given the schedule of reinforcement and the schedule of slide presentation, a pigeon could earn between 0 and 3 reinforcements by pecking in the presence of a tree slide. In the presence of non-tree slides, pecking earned no reinforcement. Each session showed the same 80 slides in a new random order. (See Herrnstein (1979) for further procedural details).

The discrimination between tree and non-tree slides formed about as rapidly as possible for three of the four pigeons in the experiment, and the fourth was not much slower. In figure 1, the acquisition curves are in terms of mean rank of positive slides, r_p, obtained by ranking the rates of pecking in the presence of all 80 slides, then averaging the ranks earned by just the

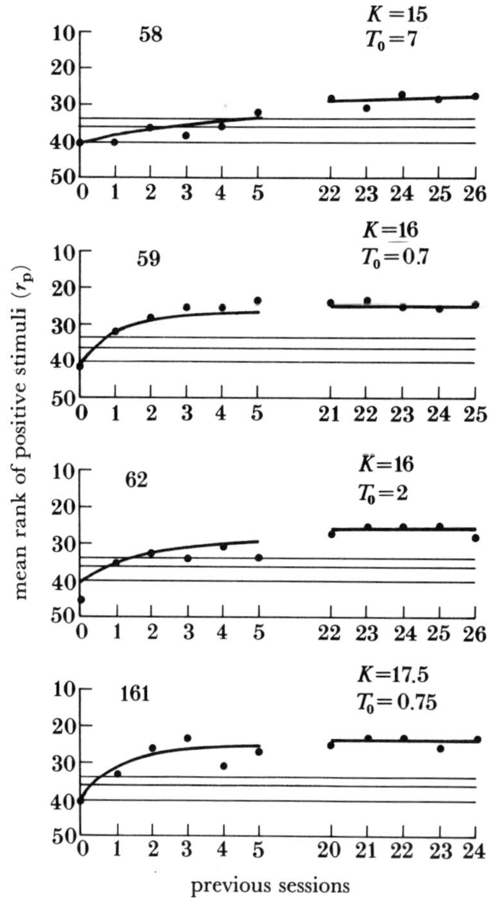

FIGURE 1. Four pigeons were reinforced for pecking in the presence of 35 mm slides of natural scenes containing trees but not in the presence of those lacking trees. Rates of responding to the 80 slides were ranked for each session; the ordinate gives the average rank earned by responding to tree slides (r_p) starting with the first session. The bottom horizontal line is at the expected value of r_p when discrimination is absent; the middle line shows r_p with discrimination at the 0.05 level of significance; the upper line, at the 0.01 level of significance. The function fitted to the points is $r_p = 40.5 - Kt/(t + T_0)$, for which the learning rate parameter is T_0 and the asymptotic level of discrimination parameter is K (from Herrnstein 1979).

tree slides. If responding is unrelated to the contents of the slides, r_p should be around 40.5, as it was at the first session (except for 62, who evidently started with a bias against pecking in the presence of tree slides). The bottom horizontal line in each panel of figure 1 is at 40.5; the two above it are at the 0.05 and 0.01 levels of discrimination. Two of the four pigeons were discriminating at or beyond the 0.01 level by the second session, having seen the slides only once before; a third (62) was discriminating at the 0.05 level by the second session; the fourth (58) took five sessions to reach and remain beyond the 0.05 level of discrimination. Even 58, the slowest pigeon, learned rapidly by the usual standards of discrimination learning in the laboratory.

Sometimes, in experiments like these, I have seen pigeons discriminating significantly before the first rotation of the slide tray has been completed. Slides never seen before are already getting sorted correctly. Something like this can be seen in this experiment as well, if the results are appropriately analysed. During the early sessions, the individual tree slides accumulated reinforcements in their presence at varying rates because of the intermittency in the reinforcement schedule, the varying presentation periods, and the pigeons' somewhat erratic initial rates of pecking. This variation gives us a chance to test whether any particular exemplar's discriminability was influenced by its reinforcement history. According to traditional interpretations of reinforcement as a means of stamping in stimulus–response connections, it would seem that some such relationship ought to have been found.

Figure 2 shows r_p for tree slides sorted by the criterion of how many reinforcements had ever been received in its presence, for individual subjects. Reinforcements are cumulated from the first session on; figure 2 shows the results for the second to the sixth session for the four pigeons. If the stamping-in theory were right, we would expect rising functions, indicating that a slide's discriminability depended on its reinforcement history. The theory is nicely illustrated by 58,

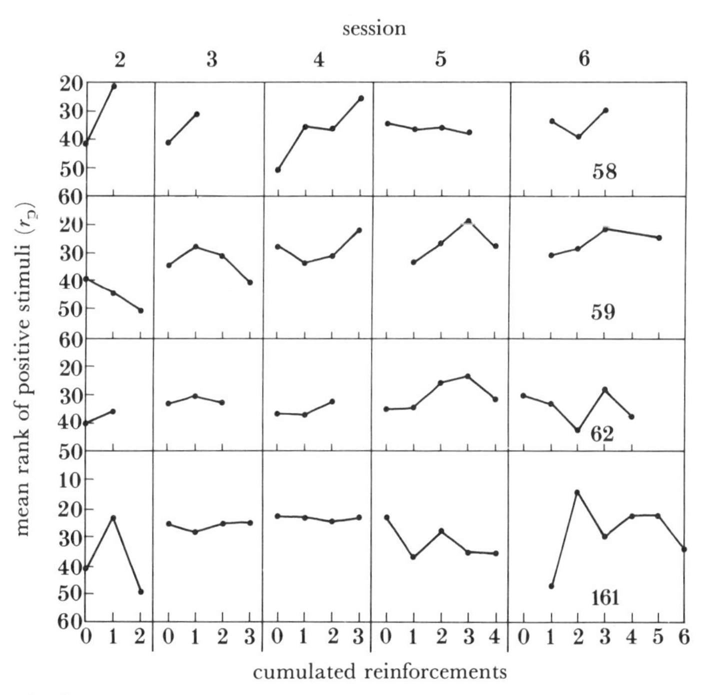

FIGURE 2. Mean rank of responding to positive stimuli classified according to the total number of reinforcements in their presence, for sessions 2–6 and for the four pigeons in the experiment (from Herrnstein 1979).

the slowest learner, for sessions 2–4, namely, before this pigeon started discriminating reliably. The three other pigeons provide little or no support for the theory. Their functions are generally horizontal, which means that even slides with no, or just 1 or 2, reinforcements in their history were being discriminated as well as those with many more reinforcements. For all four pigeons, clear discrimination coincided with evidence against the stamping-in theory. An analysis of cumulated presentation times likewise showed that discriminability was independent of prior history.

In this experiment, the same 80 slides were repeatedly shown, so that the pigeon could have solved the discrimination problem by memorizing whether each exemplar was positive or negative, never noting that one category had treeness in common. The analysis just described, however, makes that interpretation unlikely, if not insupportable, for a positive instance with no reinforcements in its history would, by hypothesis, be indistinguishable from a genuine negative instance. Yet, in most cases, they were not only distinguishable, but as distinguishable as any other positive instance. Later in this experiment, there were tests with additional tree and non-tree slides confirming that the pigeons generalized the principles used for sorting to new exemplars before any reinforcements in their presence. In other experiments (Herrnstein *et al.* 1976), the category of trees versus non-trees (and other categories) has emerged rapidly although a different set of exemplars was presented in every session. The level of discrimination with changing sets of exemplars was barely lower, if at all, than that with fixed exemplars (figure 9 in Herrnstein (1979) presents this comparison).

Generally speaking, we could describe these findings as showing that pigeons learn the relevant category as a whole, rather than just an assortment of slides. But that would be to gloss over several unanswered question. It may be asked whether the pigeon's category is literally isomorphic with the corresponding human category, for example 'tree', or whether it is, instead, some overlapping category or categories that we fail to detect. We can test particular alternative categorical principles, but not exclude the possibility of others unthought of. We cannot even be certain that the pigeons see the slides as representing objects in a three-dimensional space or as flat shapes on a plane. In another terminology, the pigeon *seems* to be displaying object constancy when it recognizes a tree in a new slide but it may only be responding to a two-dimensional form that resembles previously reinforced two-dimensional forms. I have no answer to these questions, only some guesses offered later.

Concentrating on experiments with pigeons looking for trees may foster groundless hypotheses. For example, pigeons may be thought to have a special talent for visual categorization because some breeds are capable of homing, a task calling for considerable sensory powers. Or, trees are found in the pigeons' natural habitat, so this may be supposed to be a category with special salience for them. In fact, neither pigeons nor trees, nor their coincidence, seem to be very special as far as categorical powers are concerned. Alan Kamil and his associates at the University of Massachusetts have found that blue jays are highly skilled at categorizing photographs containing particular kinds of moths, or silhouettes of particular kinds of partially eaten leaves, depending on which kind of caterpillar had eaten it (Real *et al.* 1985; Pietrewicz & Kamil 1977, 1981). Turney (1982) has shown that mynah birds can recognize trees or people in pictures, and signal their identification by vocal utterances – words, in short – instead of pecking at buttons. A parrot has learned to answer, vocally, questions about objects' shapes and colours, even about objects not seen before (Pepperberg 1983). A sizeable literature (reviewed by Herrnstein 1984) has established at least as much categorical capacity in monkeys and apes as in pigeons. Until recently it has simply been assumed that discrimination by subhuman animals is always to be explained in terms of simple generalization from specific exemplars. Despite the universality of that assumption, no species has yet been shown to be that limited in its categorical capacities, as far as I know. Each new effort to push the limits of animals' conceptual abilities seems to have a good chance of finding new abilities. The limits to date are more in the experiments tried with animals than in the animals.

Nor are species limited to the categories of their natural habitats. Peter de Villiers and I found quite good performance by pigeons looking for fish versus non-fish in underwater photographs (Herrnstein & de Villiers 1980). The ancestors of today's pigeons and fish have not shared an environment for tens of millions of years; the latter is not part of the former's natural habitat. Michael Morgan and his associates, then at Cambridge University, first showed pigeons capable of discriminating alphanumeric characters in essentially any typeface (Morgan *et al.* 1976). That finding has now been confirmed and extended by Donald Blough (submitted for publication), who also showed that the errors pigeons make in discriminating letters from each other can be explained by approximately the same underlying feature structure as are human judgments of letter similarity. Delius & Nowak (1982) have demonstrated that pigeons can recognize symmetry *per se* in visual patterns, to some extent independently of the shapes with which they are initially trained. Fish, letters, and symmetry are about as discriminable for pigeons as trees and people, so categorization cannot be limited to the objects of natural habitats.

The basic research strategy in categorization experiments on animals has been to establish a discrimination with a set of exemplars and then to see to what extent it generalizes to new exemplars. If it generalizes substantially, the argument goes, then we can infer some principle of extension. The ambiguities arise because there are always multiple principles that could encompass the observed extension, not all of which are equally interesting. A pigeon finding a specific woman in photographs, for example, is a more arresting result than one finding flesh-toned blobs of a particular shape on a plane surface (Herrnstein *et al.* 1976). The latter finding would be thought of as explaining away anything surprising about the former finding. It would be interpreted as evidence against the claim that the pigeon recognized a particular person in favour of the more mundane claim that a set of arbitrary two-dimensional patterns happen to resemble each other. How can we distinguish between claims of these two types when an animal learns to generalize to new exemplars?

When we hold constant a principle of categorization and vary exemplars, we are hoping that the lines of generalization will converge on something useful or illuminating. It is possible, however, to reverse the strategy, to hold the exemplars constant and separately vary the principle. In a new experiment by William Vaughan and me, just that new strategy has been attempted. Our argument is that, to the extent that we can dissociate a principle of categorization from the particular exemplars that convey it, we are adding credence to the notion that more is involved than mere similarity (see Lea (1984) for a fuller statement of this point).

In this experiment, pigeons control the advance switch of a slide projector by pecking a response key whenever they want a new slide. In a single session, they could advance all around an 80-slide carousel plus part of a second time around, but the results I will illustrate are from the first rotation in any session. The 80 slides consisted of 40 with trees and 40 without, inserted into the tray in a different random order for every session. Tree slides signalled one schedule of reinforcement; non-tree slides, another. By advancing the tray, the pigeon would shift from one reinforcement schedule to the other if the slide change was also a change from a tree to a non-tree or vice versa. If the next slide was in the same category, as it would be half the time, the schedule would not shift, but the pigeon could change the slide again almost immediately.

The schedules of reinforcement constituted a modified concurrent variable-interval, variable-interval, a much studied procedure in research on choice (Herrnstein 1970; de Villiers 1977; Herrnstein & Vaughan 1980). The procedure fixes a certain ratio of reinforcement rates for the two alternatives (that is, tree versus non-tree) to which animals tend to match their ratio of times allocated to each, a relationship known as the matching law. During the course of the experiment, we fixed the ratio of reinforcement rates at 1:1, 5:1, 1:3, 3:1, and 1:5 for trees and non-trees, respectively. In addition to the changing reinforcement ratios, five different sets of 80 slides were used, each one containing 40 trees and 40 non-trees.

How the reinforcement ratios and the different slide sets were programmed is shown in figure 3, along with the major results for the four pigeons in the experiment. The dots show relative reinforcement rates for successive experimental sessions starting with the first. For the first 21 sessions, the value hovers around 0.5; the reinforcement rate ratio was 1:1. Then, the ratio goes to 5:1, and the relative reinforcement rate rises to the vicinity of 0.833 until session 52, after which it drops to 0.25 as the reinforcement rate ratio fell to 1:3, and so on.

The first set of slides was used for the first 14 sessions, then at '2', the second set replaced it. Note that the schedule of reinforcement did not change at this point, but remained at 1:1 until session 21. When the schedule changed to 5:1 (at session 22), the second slide set continued to be used. It was replaced at session 40, while the reinforcement schedule stayed at 5:1. In this way, changes in the slide set and the reinforcement schedule alternated. The last change was on session 157, when the fifth slide set was replaced by the first set.

The solid lines in figure 3 trace the pigeons' behaviour as a proportion of time spent in the presence of tree slides. The matching law implies that this proportion should follow the relative reinforcement points, and it approximately does. The small but systematic deviation of the line towards 0.5, a phenomenon called 'undermatching', need not concern us here (but see Baum 1974, 1979; Wearden 1981). More to the point is the evidence that at each reinforcement ratio, the behavioural proportion transferred from one set of slides to another. Occasionally, the change in slide set caused a slight and transient regression of performance toward 0.5, the expected level if discrimination had been obliterated by the slide change. However, even in the worst case, transfer was almost complete and immediate. Figure 3 itself shows that it takes a pigeon a few sessions to reach stable performance when reinforcement schedules are changed; no such new acquisition was observed when the slide sets were replaced. The first four slide set changes (at sessions 15, 40, 81, and 120) show that an existing discrimination in performance transfers along the lines of the tree–non-tree categorization. The fifth slide set change tells us that plus something more.

On session 157, the fifth slide set was replaced by the first set as the pigeons were spending about 0.20 of their time with tree slides. The behavioural proportion transferred with almost no perturbation. Slide set 1, however, was previously seen with a reinforcement ratio of 1:1 and a behavioural proportion close to 0.5. If the pigeons had learned specific exemplars rather than a general category, the change on session 157 should have caused a return to approximately equal time allocated to trees and non-trees, which was the categorical discrimination when slide set 1 had last been seen. The lack of a change in behaviour on session

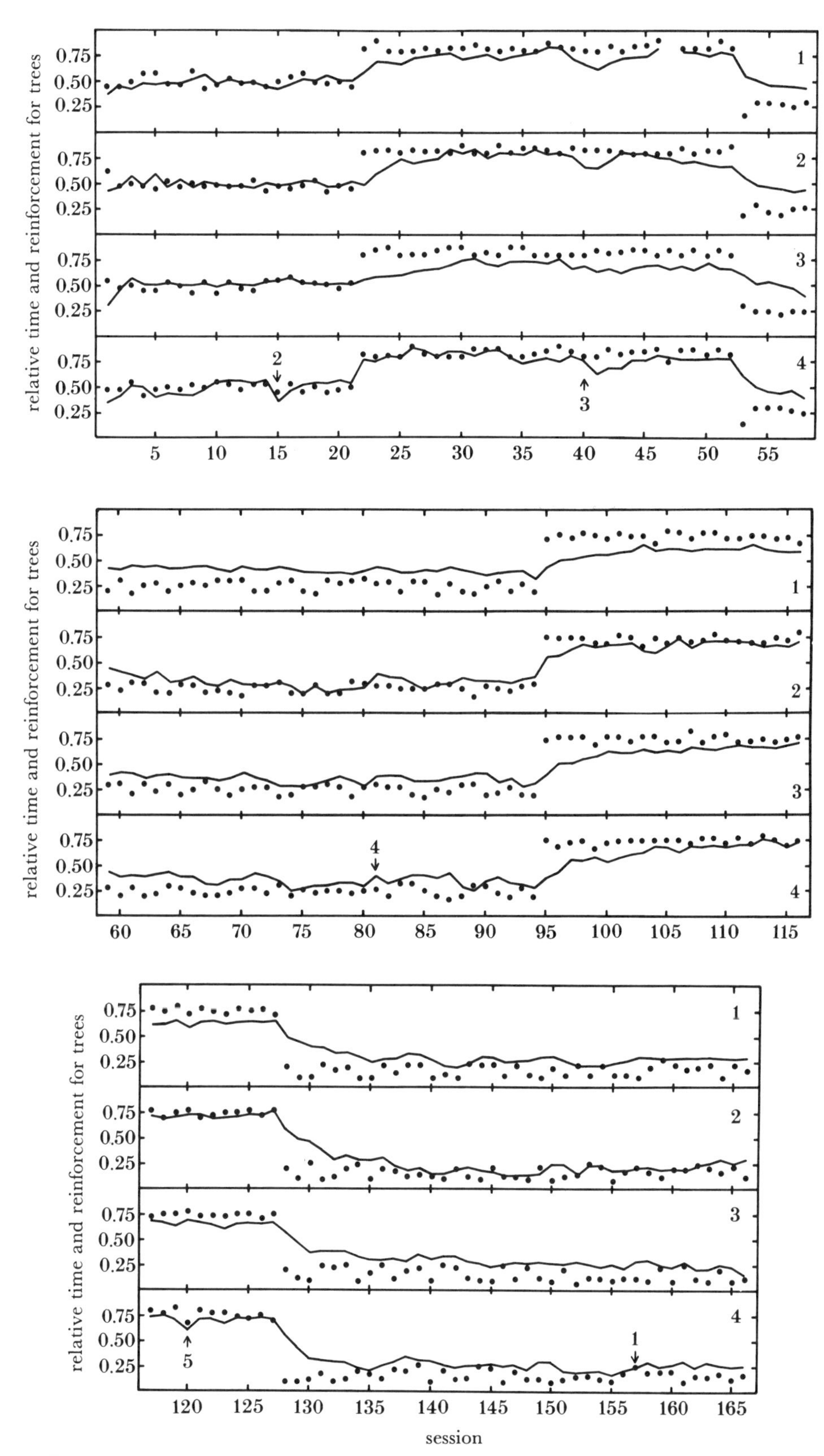

FIGURE 3. Four pigeons worked on a modified concurrent variable-interval, variable-interval schedule: one rate of reinforcement was signalled by slides containing trees and the other by slides not containing them. The pigeon controlled the advance switch of the slide projector. Relative reinforcement rate, which was changed four times during the experiment, is plotted by filled circles; relative time spent in the presence of trees is given by the continuous line. At the numbered arrows (2, 3, 4, 5 and 1), the prevailing set of 80 slides was replaced by another set. At the arrow labelled '1', the fifth slide set was replaced by the one used for the first 14 sessions of the experiment. (This experiment was performed in collaboration with W. Vaughan, Jr.)

157 says that a new discrimination had transferred to an old set of exemplars. While we still cannot say just what the pigeons are using as the criteria for discriminating one class of slides from the other, we may safely say the criteria are more general than a mere aggregation of exemplars. A bit of anthropomorphism may help. At the fifth change in the slide set, the pigeon seemed to be guided by how they were currently responding to 'trees' and 'non-trees', rather than how they had previously learned to respond to this particular slide set. The next section presents evidence that this lack of regression to an earlier performance was not due to the pigeon's inability to remember earlier discriminations.

2. Learning exemplars

Having made the case that animals can form categories that transcend mere exemplars, this section concerns the pigeon's quite striking powers of exemplar learning and retention, as revealed in a new series of experiments by William Vaughan and Sharon Greene (Greene 1984; Vaughan & Greene 1984*a*, *b*).

Instead of requiring pigeons to discriminate between collections of slides differing in some categorical respect, Vaughan and Greene have used collections of slides in which 'correct' and 'incorrect' exemplars are chosen arbitrarily. For example, 80 slides depicting outdoor scenes around the laboratory at Cambridge, Massachusetts, were arbitrarily divided into positive and negative categories and shown to three pigeons in a different random order in every session. The only way performance could improve was by memorizing whether each exemplar was positive or negative, since there was no general rule that divided the two categories. Figure 4, at A, plots the acquisition of the discrimination by increasing values of a statistic, ρ, which expresses the probability that a positive exemplar will be ranked higher than a negative (Herrnstein *et al.* 1976; Herrnstein 1979). When ρ is 0.5, discrimination is absent; when it passes 0.61, discrimination is significant at the 0.05 level. At $\rho = 0.95$, the significance level is well beyond 10^{-10}. Acquisition was not much slower than in figure 1, where tree and non-tree slides were being discriminated, but such a comparison must be qualified by the differences in procedural details, including two rotations of the carousel tray per session for this experiment versus only one for the other.

At B, C, and D in figure 4, new trays of 80 slides were introduced, each time with an arbitrary separation into positive and negative categories. To a human observer, the four sets of slides looked like random assortments of local scenes. The sessions omitted in figure 4 included a variety of tests for possible artefacts (described in Vaughan & Greene 1984*a*), all of which were negative. Each set was rapidly learned to a high level. Then, at E, the eight sessions cycled through the four sets twice (4, 3, 2, 1, 4, 3, 2, 1 in consecutive sessions). The pigeons were evidently able without difficulty to tell whether virtually all of 320 slides fell into one or the other of two arbitrary categories. Retention of these discriminations survived without further training for over a year, though with some erosion.

Vaughan and Greene also showed that the pigeons generalized what they knew about slides to their mirror reversals: putting a slide into the tray backwards did not much disrupt its discriminability. (Up–down inversions, in contrast, were highly disruptive.) But this does not mean that left–right reversal is absent from the pigeon's internal representation of a slide, as Vaughan and Greene proved in a subsequent procedure. If slides repeatedly reversed their status as positive or negative depending on their left–right orientation, the pigeons learned that too.

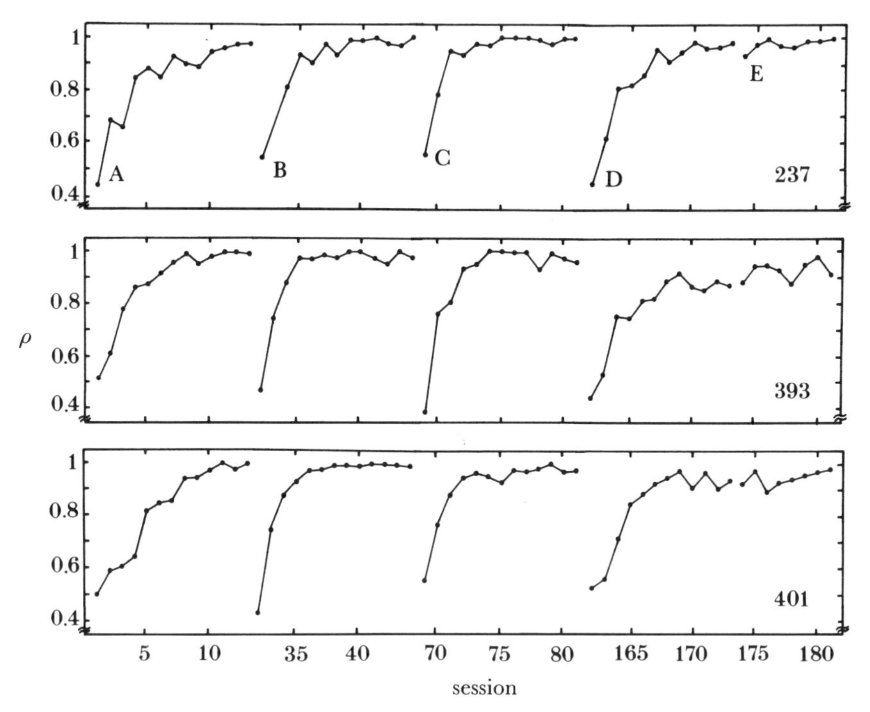

FIGURE 4. Four pigeons learning to sort sets of 80 slides of natural scenes into two arbitrary categories. The measure of discrimination, ρ, estimates the probability of ranking a positive instance over a negative instance (see Herrnstein (1979) for the relation between this measure, r_p, and signal detection analysis). Without discrimination, ρ approximates 0.5; discrimination achieves conventional statistical significance at ρ of about 0.61. At B, C, and D, new sets of 80 slides were introduced. At E, each session used one of the four slide sets, starting with the last and cycling back to the first twice: that is, in the order 4, 3, 2, 1, 4, 3, 2, 1 (from Vaughan & Greene 1984*b*).

They then responded to a slide in one orientation but not to its reversal, just as if the two orientations were two slides. A pigeon *can* distinguish mirror reversals, but it *will* only if the contingency of reinforcement requires it to do so.

The final finding to be reviewed here was a serendipitous discovery by Sharon Greene (Greene 1984). She was interested in testing whether pigeons could master relational concepts, as contrasted with the object concepts described earlier. In one experiment, pigeons looked at blocks of 10 consecutive slides composed of four slides shown twice each and two slides shown only once, irregularly interspersed with each other. Responding was reinforced only in the presence of slides being seen for the first time in a block. The second time, the slide was negative. A given set of 10 slides was used again and again to see if repetition within the identical pairs would become a basis for descrimination. Repetition prescribes a relation between stimuli, rather than a set of characteristics defined over individual exemplars, such as the class of trees or fish. The concept of repetition appeared to be rapidly learned by the pigeons, with responding largely confined to first presentations of a slide within a block.

On closer scrutiny, however, Greene discovered that it was not repetition alone, or for some pigeons, not repetition at all, that was controlling behaviour. To obtain the multiple copies of slides, Greene had photographed scenes several times in rapid succession, taking care to try to keep the camera steady. Then, in the experiment, for pairs of presumably identical slide pairs, one photograph was always the first and the other, the second. The slide pairs looked identical to casual inspection, but if they were not, then the pigeon might have learned to respond to the difference between them. This possibility was tested quite simply by reversing the order of presentation within slide pairs, and the result was a drastic, in some instances total, disruption of discrimination. Impelled by this result, Greene searched her slides and found minute differences in shadows or in the edges of the scene that could have been the basis of discrimination.

In the next phase of the experiment, Greene randomized presentation order within slide pairs. Now, only repetition could provide a reliable signal for non-reinforcement. The pigeons relearned the discrimination, although not to the level obtained when other cues were being used. Pigeons could, in short, learn a relational principle, but apparently found it less congenial than to respond to what, to a human observer, would seem to be obscure features of the slides. Just as for mirror reversal, the relational concept of repetition needed selective reinforcement before it took control of behaviour.

3. Pigeon intelligence

Pigeons can categorize at levels of abstraction that defy both explanation and simulation, but they do not have to. They also have surprisingly strong resources for learning specific exemplars. Vaughan and Greene's data on exemplar retention are reminiscent of the remarkable ability of some animals to recall where they hide bits of food. Vander Wall estimates that a particular species of nutcracker caches tens of thousands of seeds in a season, then recovers well over 50% of them on the basis of the specific cues in the vicinity of the cache (Vander Wall 1982). The homing behaviour of pigeons is yet another case of notable exemplar recognition.

In these respects, pigeons display surprising cognitive capacity. In other respects, they live up to a reputation for dullness. One form of discrimination at which pigeons, and perhaps other animals, are deficient is the relational. Greene's repetition experiment suggests a degree of reluctance to use a relation as a basis for discriminating. Similarly, it has proved to be difficult, though not impossible, to teach pigeons to generalize the concept of sameness or difference (Wright *et al.* 1984; Zentall *et al.* 1984). Pigeons learn only with some trouble that they are being reinforced for responding to a stimulus that matches another stimulus in some respect, or that does not match it. When they do learn to solve such problems, they do so with less generality than they could, as if the abstract relation of identity or difference is contaminated by the specific stimulus features of the stimuli used to illustrate it (Lombardi *et al.* 1984). There is also some question whether they can recognize the relation – identity or difference – as such, above and beyond applying it to a series of test stimuli (Premack 1976, 1978).

In Blough's letter discrimination experiment, the pigeons were required to peck at a particular letter out of three presented. For various procedural reasons, the correct letter was always pitted against two identical incorrect letters: an A versus two Bs or a K versus two Ds, for example. The pigeons could have solved the problem by always pecking the odd letter, but they did not. The relational solution was apparently harder for them than the absolute one: learning to peck the A, then the K, and so on through the alphabet. What could have been reduced to the learning of one relational concept remained for the pigeons the learning of 26 separate letters.

Even in the tree–non-tree kind of discrimination, pigeons may betray a lack of relational insight. A two-way categorization involves just one bit of information: if we can recognize trees, then non-trees are simply the complement. Logic would dictate symmetrical errors, once the biases due to reinforcement are taken into account. Yet, pigeons are more likely to label a non-tree a tree than vice versa, whether trees are the reinforced or the non-reinforced category (Herrnstein 1979). It is as if the pigeon is trying to formulate two sets of rules, one for trees and another for non-trees. The latter is bound to be harder.

In our laboratory, we tried, without much success, to demonstrate mental rotation by pigeons, using adaptations of procedures developed for human subjects by Roger Shepard and his associates (for example, Shepard & Metzler 1971). Hollard and Delius have, in contrast, reported that pigeons can learn to match a simple geometrical figure to its duplicate or to its mirror reversal, and to generalize the skills across varying rotations in the picture plane of the figures (Hollard & Delius 1982). There are, however, grounds for scepticism about whether the pigeons are, in fact, rotating the figures mentally, rather than, say, rotating their heads or learning to identify distinguishing features of rotated or mirror-reversed figures. The main reason for the scepticism is that the pigeon data lack the reaction time function considered to be the *sine qua non* of mental rotation in humans. For humans, the time to decide whether two figures are merely rotations of each other or rotations plus mirror reversals is a linear function of their angular rotation relative to each other, as it was for the human subjects that Hollard and Delius also tested. For their pigeons, however, response latency was independent of angular rotation, and was, moreover, considerably shorter than the human latencies.

Hollard and Delius suggest that the pigeon has a special evolutionary adaptation for assessing rotational invariances by parallel processing, while human observers are limited to a slower, rotation-dependent serial process. While the suggestion is certainly interesting, it has only the data in their own experiment to support it. Our own failure to find evidence of mental rotation in pigeons, combined with the flat functions in Hollard and Delius's experiment, may indicate a deficiency in the pigeon's ability to rotate, or otherwise transform, internal representations. Hollard and Delius characterize the pigeons' performance as more efficient than the humans' because the latencies were shorter, but the pigeons had substantially more practice with the general task (90 sessions compared with 2 sessions), and their error rates were still somewhat higher.

Perhaps the most illuminating deficiency of discrimination at this point is Cerella's finding that pigeons fail to generalize across varying views of computer-generated line projections of cubes (Cerella 1977). More accurately, having seen a sector of the domain of all possible views of the cube, the pigeons generalized no more than they did in the domain of stimuli made up of the lines of the cube rotated 90° around their midpoints, a transformation that wipes out the three-dimensionality for human observers. The two sorts of stimuli are illustrated in figure 5. To a human observer, the three-dimensional invariance of the set on the left is so powerful that generalization across all views is complete and instantaneous, whereas the set on the right is just so many bunches of lines. For a pigeon, judging from Cerella's results, the set on the left is no more homogeneous than the set on the right, nor any more discriminable from stimuli outside the domain.

If there were just this finding of Cerella's, we might conclude that pigeons cannot see three-dimensional objects in plane projection. Some may believe that that is true, but they would be hard put to explain how pigeons recognize trees, people, fish, and so on, in photographs and generalize so readily to new exemplars. Another possibility is that the problem

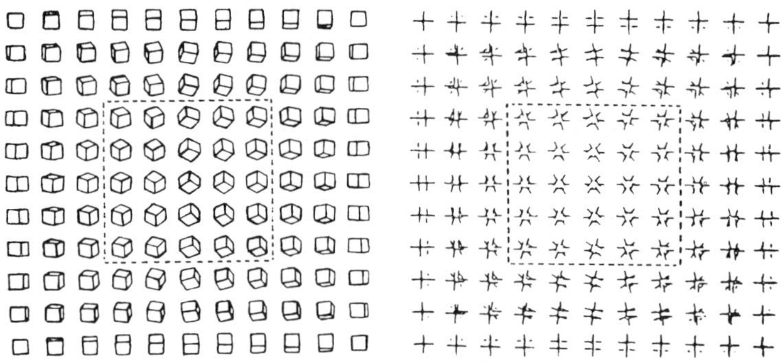

FIGURE 5. Pigeons learned to respond to stimuli drawn from the region inside the dotted squares, but not to generalize beyond this region. They generalized no more broadly with the cubes than with the line patterns (from Cerella 1977).

with the cubes is that they are too impoverished to evoke the perception of three-dimensionality. Photographs of real cubical blocks might, by this hypothesis, have attenuated the striking difference between pigeons and humans. If so, we could say that a pigeon needs more data on a plane than a human being to construct in its mind a three-dimensional isomorph, just as most ordinary people need more data than a structural engineer or an architect trained to read blueprints.

Pigeons have also failed in our laboratory to learn some categories that human observers find easy. Failed experiments are unlikely to be published, though they may contain no less information than just another demonstration of successful categorization. One reason they do not get published is that the experimenter gets discouraged before pinning down the failure to produce discrimination. We gave up trying to teach pigeons to classify photographs of bottles versus non-bottles, chairs versus non-chairs, wheeled vehicles versus anything that was not a wheeled vehicle. We also failed with photographs of the small plastic cups from which the pigeons in the experiment ate their daily rations versus photographs of other scenes around the laboratory. Not every pigeon was a total failure in these experiments; often, a few subjects would learn to recognize a few particular views of target objects. Nevertheless, it was clear that the level of categorization fell far short of that with other objects, such as trees or fish.

It may be helpful to try to sort through the pigeon's failures in categorization. The pigeon's apparent deficiencies with visual classification, taking the human observer as the norm, fall into four general areas, as follows.

(i) Pigeons fail with certain object categories. To date, these cases are all with man-made and three-dimensional objects, for example, bottles, as distinguished from fish (not man-made) or letters (not three-dimensional). If the pattern holds up, it may be because the construction of a mental, three-dimensional invariant places greater cognitive demands than two-dimensional invariants. Man-made objects, reflecting functional and other constraints peculiar to the human context, may simply tax the pigeon's capacities more than objects forged under the same constraints as those of the pigeon's environment. This is different from saying that man-made objects are unfamiliar to the pigeon. Rather, it is to say that the appearances of objects depend in some measure on the processes that give rise to them, and that some of those processes for man-made objects are not ones for which the pigeon's visual system has evolved constancies.

(ii) Along similar lines, pigeons appear to need, not only different, but more information than human observers to construct a three-dimensional image from a plane representation, as in Cerella's cube experiment. Other experiments (Cerella 1980, 1982; discussed in Herrnstein 1984) similarly suggest that line drawings that evoke three-dimensional images for humans fail to do so for pigeons.

(iii) The lesser capacity for constructing three-dimensional percepts may be part of a more general deficiency in mental manipulation. The absence of mental rotation may be an illustration, and so may the breakdown in discrimination for inverted images (Vaughan & Greene 1984*a*). There have been few tests of how readily pigeons compensate mentally for other stimulus transformations, such as changes in size, colour, angle of regard, etc. It may seem reasonable that such compensations must be involved in the experiments showing object constancy in natural settings, but the photographs in those experiments are usually so rich and redundant that substantial deficits could escape detection.

(iv) Relational principles, for example, oddity, identity, complementarity, repetition, characteristically involve not just classes, but classes of classes. To be able to choose the odd stimulus in a trio, for example, requires not only an identification of the three objects, but a formal or logical structure across the objects. Pigeons can solve such two-tiered categorization problems, but they often seem disposed to stay at the pre-logical level. Even when the reinforcement contingencies drive them to the logical level, the performance often shows traces of contamination by the lower level. The logical solution tends not to generalize as far from the specific stimuli used in training as it should. This is sometimes thought of as the pigeon's over-reaction to 'novelty', but it is novelty of a special kind. What the pigeon appears to have difficulty with is the isolation of a relationship among variables, as contrasted with a representation of a set of exemplars.

This last deficiency may be just another instance of the pigeon's weakness in mental manipulation, but it is an instance with special significance. A facility with relationships among variables, free of contamination by specific values, is what gives both the computer and human language their power. Perceptual constancies allow pigeons and other animals to navigate in a world of objects, but the reluctance, if not the inability, to manipulate mentally the abstract relationships into which objects may be placed may be what prevents animal thought from dealing productively with the world represented internally.

4. A comment on method

What an animal *can* do may differ from what it *does* do on any occasion, as I noted above. This distinction is a corollary of the environment's control over psychological processes, as illustrated by the power of reinforcement. An animal that fails to perform in a particular way may indeed lack the necessary capacity or it may just be that the particular test has failed to activate it. The animal's psychology is always linked to some context, which is a source of adaptiveness and also a hazard to students of the animal mind. In his wonderfully evocative book, *Animal thinking*, Donald Griffin comments about the oddity problem: 'chimpanzees... have learned to generalize oddity as such... pigeons have much greater difficulty with comparable problems, but do better than cats and raccoons' (1984, p. 140). It is at least equally plausible, in my judgment, that cats and raccoons would do better than pigeons if they had better reasons for doing so than they did in the tests at issue. Even the comparison between chimpanzee and pigeon should be qualified by the limitations of the procedures used. Comparative psychology is a far more elusive subject than, say, comparative anatomy, for its facts are peculiarly dependent on its techniques. It is not impossible to make sound comparisons across species, but it is difficult.

That reservation applies to my comparisons, as well as Griffin's and others. At most, we may conclude that the data suggest that pigeons fail to manipulate internal representations or to use logical structures in situations in which human subjects would not fail to do so. This is different from saying, for example, that people use logical relations and pigeons do not. For one thing, pigeons can be induced to behave logically, and, for another, there is no lack of evidence for human illogicality: halo effects, logical fallacies, 'cognitive illusions', and false inferences have inspired much research on human thinking. If there are different upper bounds for people and pigeons in logical powers or mental manipulativeness, as there no doubt are, finding them would require knowing how to push each species to its limit. Interesting as it may be, performance at the limit would be of questionable relevance to behaviour in natural settings, which is rarely driven to any sort of limit. If making sense of performance in natural settings is one of the riddles of natural categorization, solving it means finding out how the ordinary contingencies of reinforcement interact with a species' perceptual and behavioural predispositions.

Preparation of the article, as well as the conduct of several of the experiments reported here, were supported by grant no. IST-81-0040 from the National Science Foundation to Harvard University. Many thanks are due to the organizers of the conference for which the article was prepared and to W. Vaughan for permission to present previously unpublished results from an experiment we conducted in collaboration.

References

Baum, W. M. 1974 On two types of deviation from the matching law: Bias and undermatching. *J. exp. Analys. Behav.* **22**, 231–242.

Baum, W. M. 1979 Matching, undermatching, and overmatching in studies of choice. *J. exp. Analys. Behav.* **32**, 269–281.

Blough, D. S. 1985 Discrimination of letters and random dot patterns by pigeons and humans. (Submitted.)

Cerella, J. 1977 Absence of perspective processing in the pigeon. *Pattern Recog.* **9**, 65–68.

Cerella, J. 1980 The pigeon's analysis of pictures. *Pattern Recog.* **12**, 1–6.

Cerella, J. 1982 Mechanisms of concept formation in the pigeon. In *Analysis of visual behavior* (ed. D. J. Ingle, M. A. Goodale & R. J. W. Mansfield), pp. 241–263. Cambridge, Mass.: M.I.T. Press.

Delius, J. D. & Nowak, B. 1982 Visual symmetry recognition by pigeons. *Psychol. Res.* **44**, 199–212.

de Villiers, P. 1977 Choice in concurrent schedules and a quantitative formulation of the law of effect. In *Handbook of operant behavior* (ed. W. K. Honig & J. E. R. Staddon), pp. 233–287. Englewood Cliffs, N.J.: Prentice-Hall.

Greene, S. L. 1984 Feature memorization in pigeon concept formation. In *Quantitative analyses of behavior, discrimination processes* (ed. M. L. Commons, R. J. Herrnstein & A. R. Wagner), pp. 209–231. Cambridge, Mass.: Ballinger.

Griffin, D. R. 1984 *Animal thinking*. Cambridge, Mass.: Harvard.

Herrnstein, R. J. 1970 On the law of effect. *J. exp. Analys. Behav.* **13**, 243–266.

Herrnstein, R. J. 1979 Acquisition, generalization, and discrimination reversal of a natural concept. *J. exp. Psychol.: Anim. Behav. Process.* **5**, 116–129.

Herrnstein, R. J. 1984 Objects, categories, and discriminative stimuli. In *Animal cognition* (ed. H. L. Roitblat, T. G. Bever & H. S. Terrace), pp. 233–261. Hillsdale, N.J.: Erlbaum.

Herrnstein, R. J., Loveland, D. H. & Cable, C. 1976 Natural concepts in pigeons. *J. exp. Psychol.: Anim. Behav. Process.* **2**, 285–311.

Herrnstein, R. J. & de Villiers, P. A. 1980 Fish as a natural category for people and pigeons. In *The psychology of learning and motivation* (ed. G. H. Bower), vol. 14, pp. 60–97. New York: Academic Press.

Herrnstein, R. J. & Vaughan, W., Jr 1980 Melioration and behavioral allocation. In *Limits to action* (ed. J. E. R. Staddon), pp. 143–176, London: Academic Press.

Hollard, V. D. & Delius, J. D. 1982 Rotational invariance in visual pattern recognition by pigeons and humans. *Science, Wash.* **218**, 804–806.

Lea, S. E. G. 1984 In what sense do pigeons learn concepts? In *Animal cognition* (ed. H. L. Roitblat, T. G. Bever & H. S. Terrace), pp. 263–277. Hillsdale, N.J.: Erlbaum.

Lombardi, C. M., Fachinelli, C. C. & Delius, J. D. 1984 Oddity of visual patterns conceptualized by pigeons. *Anim. Learn. Behav.* (In the press.)

Morgan, M. J., Fitch, M. D., Holman, J. G. & Lea, S. E. G. 1976 Pigeons learn the concept of an 'A'. *Perception* **5**, 57–66.

Pepperberg, I. M. 1983 Cognition in the African Grey parrot: Preliminary evidence for auditory/vocal comprehension of the class concept. *Anim. Learn. Behav.* **11**, 179–185.

Pietrewicz, A. T. & Kamil, A. C. 1977 Visual detection of cryptic prey by bluejays (*Cyanocitta cristata*). *Science, Wash.* **195**, 580–582.

Pietrewicz, A. T. & Kamil, A. C. 1981 Search images and the detection of cryptic prey: An operant approach. In *Foraging behavior: ecological, ethological and psychological approaches* (ed. A. C. Kamil & T. D. Sargent), pp. 311–331. New York: Garland Press.

Premack, D. 1976 *Intelligence in ape and Man.* Hillsdale, N. J.: Erlbaum.

Premack, D. 1978 On the abstractness of human concepts: why it would be difficult to talk to a pigeon. In *Cognitive processes in animal behavior* (ed. S. H. Hulse, H. Fowler & W. K. Honig), pp. 423–453. Hillsdale, N. J.: Erlbaum.

Real, P. G., Iannazzi, R., Kamil, A. C. & Heinrich, B. 1985 Discrimination and generalization of leaf damage by blue jays (*Cyanocitta cristata*). (Submitted.)

Shepard, R. N. & Metzler, J. 1971 Mental rotation of three-dimensional objects. *Science, Wash.* **171**, 701–703.

Turney, T. H. 1982 The association of visual concepts and imitative vocalizations in the mynah (*Gracula religiosa*). *Bull. Psychon. Soc.* **19**, 56–62.

Vander Wall, S. B. 1982 An experimental analysis of cache recovery in Clark's nutcracker. *Anim. Behav.* **30**, 84–94.

Vaughan, W., Jr & Greene, S. L. 1984*a* Acquisition of absolute discrimination in pigeons. In *Quantitative analyses of behavior, discrimination processes* (ed. M. L. Commons, R. J. Herrnstein & A. R. Wagner), pp. 231–239. Cambridge, Mass.: Ballinger.

Vaughan, W., Jr & Greene, S. L. 1984*b* Pigeon visual memory capacity. *J. exp. Psychol.: Anim. Behav. Process.* **10**, 256–271.

Wearden, J. H. 1981 Bias and undermatching: Implications for Herrnstein's equation. *Behav. Analys. Lett.* **1**, 177–185.

Wright, A. A., Santiago, H. C., Urcuioli, P. J. & Sands, S. F. 1983 Monkey and pigeon acquisition of same/different concept using pictorial stimuli. In *Quantitative analyses of behavior, discrimination processes* (ed. M. L. Commons, R. J. Herrnstein & A. R. Wagner), pp. 295–319. Cambridge, Mass.: Ballinger.

Zentall, T., Edwards, C. A. & Hogan, D. E. 1983 Pigeons' use of identity. In *Quantitative analyses of behavior, discrimination processes* (ed. M. L. Commons, R. J. Herrnstein & A. R. Wagner), pp. 273–295. Cambridge, Mass.: Ballinger.

Discussion

D. I. Perrett (*University of St Andrews, U.K.*). My question arises from work (done at St Andrews) on perceptual categorization of faces by macaque monkeys. In the temporal cortex we have found substantial numbers of cells that respond only to the sight of faces. These selective responses generalize over changes in view of faces that are observed with differing backgrounds, distance, orientation and lighting. For some cells selectivity to the general characteristics of faces and even to the identity of particular faces is found to be based on sensitivity to distinctive features of that face and for other cells selectivity is based on the combination or configuration of features. At the cellular level categorization was thus found to be based on key features or their configuration.

At the behavioural level we have trained monkeys to discriminate real faces and picture faces (of monkeys, humans and animals) from objects or pictures not containing faces (unpublished studies). Two monkeys trained in this task correctly categorized 90% of novel pictures on the first trial of their presentation. Generalization tests involving jumbled faces and different regions of the face presented in isolation, revealed that the monkeys based discrimination performance on the presence of key features (such as an eye), since individual facial features and jumbles were treated as faces. It was only after explicit training of discrimination between normal faces and jumbled face features that we were able to show that the monkeys could utilize the configurational cues present in faces. Indeed in this latter task monkeys behaved similarly to humans and showed an increased reaction time to categorize faces as normal when the faces were inverted.

My question is, therefore, what is the visual basis of the pigeon's performance? To be more precise, do generalization tests of the above type reveal that pigeons are attending to characteristic features or configurational cues, or both?

R. J. HERRNSTEIN. For the kinds of photographic stimuli described in my article, there has been no systematic attempt to contrast control by features versus control by general configuration. The impediments to research on the issue are both technical and conceptual. Consider, for example, a discrimination between photos containing trees and those not containing them. Many photographs contain partial views of trees, because of obstructing objects or the edges of the picture. The pigeons generally respond to these as trees, just as a human observer would. But we cannot interpret this response to a partial tree, by either pigeons or people, as evidence against a configurational category, for a partly obstructed tree in a natural setting is in some sense just as good an instance of a tree as an unobstructed one.

It may seem that what is needed are pictures of natural scenes with unnaturally fragmented trees in them. Imagine, for example, a scene in which the parts of a tree have been randomly dispersed in all directions. Would the pigeon respond to this as a tree? The experiment has not been done, and, with the simple slide projectors we work with, it would be quite a hard experiment to do. With computer-driven visual displays, such experiments may be just a few years away.

In my opinion, such experiments are well worth doing even though they may tell us less than some may think about whether the categorization is featural or configurational. Suppose the pigeon fails to generalize to the 'exploded' tree. This would not disprove the feature theory nor prove the configurational theory, for it is always possible that the pigeon would generalize to some other fragmentation of the tree, or that it would have generalized if original training had included more almost totally obstructed trees. Now suppose the pigeon generalizes. This does not prove the feature theory nor disprove the configurational. As Perrett's comments illustrate, animals may respond to features under one set of reinforcing conditions and shift to more global characteristics if the conditions change.

Cerella (1982) showed that pigeons, after training with normal instances, generalized to scrambled instances of the cartoon character, 'Charlie Brown'. But, from other experiments, we may safely assume that the pigeons could have been trained to discriminate between scrambled and normal 'Charlie Browns'. Cerella's finding may seem to imply featural categorization; by the same token, the hypothetical latter finding would imply that categorization is configurational. If this seeems paradoxical, it is because it tacitly assumes that the process of categorization is inflexible, rather than adapted to the contingencies of reinforcement that call it forth. As my paper suggested, what an animal does in a particular setting is not likely to be the proper measure of what it is capable of doing generally.

Reference

Cerella, J. 1982 Mechanisms of concept formation in the pigeon. In *Analysis of visual behavior* (ed. D. J. Ingle, M. A. Goodale & R. J. W. Mansfield), pp. 241–263. Cambridge, Mass.: M.I.T. Press.

Phil. Trans. R. Soc. Lond. B **308**, 145–158 (1985) [145]
Printed in Great Britain

Social foraging in marmoset monkeys and the question of intelligence

By E. W. Menzel, Jr, and C. Juno
Department of Psychology, State University of New York at Stony Brook, Stony Brook, New York, 11794, U.S.A.

A social group of five saddle-back tamarins (*Saguinus fuscicollis*) were allowed 15 min per day in a sizeable room adjacent to their home cage. Every other day two additional novel test objects were placed in the room; one contained food on first presentation, and the next day the locations of both were sometimes moved. From the outset, and even when there were 30 objects to choose from, the animals were acute in detecting the novel objects and in remembering the objects and the locations in which they had found food. Whichever individuals had eaten first were among the first to approach the next day. Subsequent tests showed that such one-trial learning was not dependent on object-novelty; that the animals probably remembered all 30 objects and the location of each; and that they spontaneously performed what amounts to generalized delayed matching to sample. The data match or surpass the asymptotic performances of other marmosets on, for example, learning set tasks but are consistent with what is known about the foraging habits of wild *S. fuscicollis*. Optimal foraging theory is less likely to be an overestimate of animals' mental capacities than previous studies are an underestimate.

1. Introduction

Taxonomists and psychologists alike have in the past often characterized marmosets (family Callitrichidae) as 'squirrel-like', 'primitive' primates. These characterizations are, however, misleading (Ford 1980; Garber 1980). Not only the distinguishing morphological features of these animals (diminuitive body size, tendency towards reproductive twinning, claw-like nails on all digits other than the hallux, and some aspects of their molar dentition) but also many features of their locomotor, postural and other behaviours and their ecological and social organization could well be relatively recent and interrelated adaptations rather than retentions of the traits of their primitive New World primate ancestors (Moynihan 1976; Szalay & Delson 1979). Ford (1980), in fact, argues that marmosets represent a clear case of phyletic dwarfing. Severely limited space and food resources, and interspecific competition for these resources, are among the most obvious candidates for the major selective forces that might have led to their dwarfing.

But regardless of how marmosets originally came to be as they are, there is no question that their foraging strategies today are quite different from those of other taxa, and from one another (Dawson 1979; Izawa 1978; Mittermeier & van Roosmalen 1981; Terborgh 1983). As Terborgh (1983, p. 94; see also Milton 1981; Richard 1981) puts it, the most important characteristic of fruits from the standpoint of their differential exploitation by primates is not their size, texture, colour, construction or taxonomic status but their characteristic concentration in space and time; and the saddle-back tamarin, *Saguinus fuscicollis*, seems to have a unique strategy, at least by contrast to the various non-*Saguinus* species in the same region, in going for fruits that occur in tiny, scattered incremental units and that furthermore ripen in a piecemeal fashion. This pattern of ripening implies that only a very small amount of food is available for eating in any given locus on any single occasion, and also that a reliable (though scanty) supply can be obtained at the same loci over a period of many weeks. Such facts, Terborgh believes, will be of the utmost significance for understanding the behavioural, social and ecological organization of *Saguinus* more generally.

Ecologically oriented students of animal learning would surely concur (for example, Johnston 1981; Kamil & Sargent 1981; Kamil & Yoerg 1982). Indeed, such facts might help to explain the sorts of laboratory data on learning and memory that we ourselves have been obtaining over the past several years. Conversely, from our data we would predict that not only are feral *S. fuscicollis* capable of foraging for the above sorts of resources in a relatively optimal fashion, but also that the mechanisms by which they do so entail exceptionally rapid learning and an ability to remember from one day to the next the visual appearances and the relative positions of many objects simultaneously. To be more specific, our data strongly suggest that at least when these animals are tested in a sufficiently 'naturalistic' fashion, they require no training beyond that which they have already picked up 'on their own' in their home environment to show performances which students of learning would construe as evidence for 'cognitive maps' (Menzel 1978; Tolman 1948) and 'win–stay, lose–shift' strategies with respect to objects and locations (Levine 1959; Kamil & Yoerg 1982; Olton 1979). Here we shall not review the results of our already-published studies (Menzel & Menzel 1979; Menzel & Juno 1982). Instead, we shall report four of our more recent tests, which extend our earlier findings, and should dispel the doubts of those students of learning set formation (Schrier & Thompson 1984) who believe that these findings do not entail anything on a par with what primates do on Harlow's (1949) classical tasks. (See also Menzel & Juno 1984.)

2. General method

The animals used in this study were five laboratory-born sibling adult *S. fuscicollis*: twin males Koni and Niko (4.5 years old), twin males Blaze and Flame (3.5 years old) and a female, Natalie (2.5 years old). They had received extensive testing in situations that involved the presentation of a single test object at a time, in precisely the same locus (for example, Menzel & Juno 1982), but this was their first test involving multiple test objects that were presented simultaneously. Their regular diet had always been given to them every day in the same food pan and usually in the same location. The largest room they had ever been in since birth was their present home cage, which consisted of a 3 m by 4.3 m by 4 m section of a slightly larger indoor room.

Inasmuch as *S. fuscicollis* typically live and forage in social groups rather than alone, we tested them as a group, simultaneously. Testing consisted of opening a guillotine door to the adjacent 6 m by 4.3 m by 4 m room and recording for a period of 15 min the precise order in which, and the time (to the nearest 15 s) at which each individual entered this test room, returned to the home cage, or approached or made physical contact with any of 30 previously designated and widely dispersed points in the test room. Further qualitative notes were also made. Observations were recorded verbally by tape recorder and timed with an electronic timer which sounded a clearly audible beep every 15 s.

Testing was conducted before the morning feeding. As soon as the 15 min 'trial' had elapsed the animals were given their customary morning rations in their home cage, and the door to the test room was closed until the next day. (Almost always they had already returned home before 15 min was up.) Each room was furnished with a network of dead trees and branches, via which the monkeys could reach virtually any point in the rooms, from floor to ceiling.

'Pretraining' consisted of allowing the animals to explore the previously novel test room 15 min per day for 10 days (by which time they seemed well habituated to the room and to the procedures) and then adding a number of small cardboard tags to serve as location markers, and giving the animals an additional two 15 min trials to habituate to them. Formal testing began the next day.

3. Test 1: reactions to novel food and non-food objects

This test was similar to those used elsewhere (Menzel 1971; Menzel & Menzel 1979, experiment 1). The most important single difference, procedurally speaking, was that some objects contained food. If *S. fuscicollis* can readily learn and remember the visual appearances and the locations and the orientations of almost any given object (Menzel & Menzel 1979), few cognitively inclined investigators would, of course, see much reason to doubt that they could also learn and remember which of these same objects had contained edibles. From this point of view, the data contain few surprises.

Method

Two new test objects were introduced into the test room every two or three days (for example, an aspirin bottle and a small bow made of red ribbon; a white plastic bag and a tin can). One object of each such pair was randomly selected as the 'positive' or food object and in it or within 25 cm of it were hidden 10 small pieces of candy. The other, 'negative', object contained no food, but (on problems 6–16) as a control for odours, the same amount of candy had been left in it for at least 10 min, and then removed just before the test began. In two problems the positive object contained food both on trial 1 and on trial 2; in all others it contained food only on trial 1. The two locations in which the objects were to be placed were selected in a non-systematic fashion from among the 30 tagged spots in the room, and then a coin-toss determined which object would be in each of these locations. In eight of the 16 problems that were thus presented over a period of 34 days the locations of the two newest objects were changed on trial 2. Twice they were directly swapped (the typical procedure in the traditional two-choice discrimination learning experiment), and the remaining six times the objects were moved to other locations, sometimes displacing older objects to a new location, and other times going themselves to new locations. Rather than discard each pair of objects after a few trials, we discarded only two objects (of which the animals remained somewhat cautious). Old objects remained in the same places from one day to the next, unless displaced by a new object. Thus, the number of test objects in the test room increased cumulatively as the experiment progressed. Unless noted otherwise, the statistical significance of the data was assessed with Monte Carlo simulation techniques.

Results and discussion

Trial 1 performance, and a general description of the animals' search strategies

As expected (cf. Menzel & Menzel 1979), the animals were from the outset very skilled at detecting the new objects on trial 1 of any given problem. They could not initially tell which of the two novel objects was 'positive' and which was 'negative'; in 7 of 16 problems they approached the latter object before they went to the former. However, only 2 of the 32 novel objects were approached later than one would expect by chance; and in 7 of the 16 problems the positive object was the first object in the room to be hit ('chance' expectancy = 1.72; $p < 0.001$), and all nine exceptions involved responding first either to the negative novel object or to an object that had been positive on a preceding problem.

In some respects the group's behaviour was very different from that seen in tests in which no objects ever contained food (Menzel & Menzel 1979). On entering the test room the animals fanned out from one another and scanned different sections of the room, as if operating as a team and actively searching for the food. They often moved at very high speeds. Koni and Natalie usually travelled together, with Koni in the lead. The number of problems in which each animal made the first object approach to be recorded on trial 1 was: Koni 10, Niko 0, Blaze 0, Flame 2, Natalie 4. The number of problems in which each was first to eat was: Koni 4, Niko 0, Blaze 2, Flame 1, Natalie 9. The last three (youngest) animals seemed by far the most reliant on cues from others, and Natalie excelled in opening the containers. Niko was typically the last to enter the room and foraged at the opposite sides of the room from Koni; in a subsequent test we removed Koni from the group and Niko immediately took over his role as 'leader'. In general, each individual followed a somewhat different strategy which was undoubtedly influenced as much by its relationships with others as by any other contingencies. Each also had at least a few locations it tended to check before anyone else.

On first detecting the food, the animals very often made a distinctive 'food call', which brought others over towards the object. (An 'alarm call' by an animal who was cautious of an object produced an even more marked group reaction. All group members typically raced over with piloerected tails and 'mobbed' the object together.) Individuals very rarely stayed at the food object until all food was gone. Typically, they grabbed a few pieces of food and ran off with it several metres. As a result, on most trials each of the five animals managed to get some food (even though any one of them would easily have consumed the total quantity that was available, if given the opportunity), and obvious agonistic behaviours towards one another were rare. Obviously, they also had to remember the locus of the food source if they were to check it out again, unless by chance they could get some cues from one another. They did clearly remember, for they usually returned to the food object after having finished their food, by-passing but ignoring other objects in the process. Other indications of their (short-term) memory, which were verified more thoroughly in subsequent tests, were the tendencies of individuals to seldom check unbaited or relatively old objects more than once or twice in a trial and to avoid checking unbaited objects that had recently been checked by a group-mate.

Learning and long-term memory

Figure 1 shows the sequence in which the first five or more objects were approached for the first time (by any animal) in each of the first 10 problems, both on trial 1 and on trial 2. (Usually most of the objects were approached at some time during the 15 min trial.) Table 1 shows the total number of 15 s intervals in which the five animals were scored as approaching or touching each of the objects on each trial that entailed no food in any object. (Each individual could receive a maximum of one count per object per interval.) It is obvious from these data that the effects of a single reinforced trial, and of relative novelty, could usually be detected for several days.

In 15 of 16 problems the first object in the room to be approached by any animal on trial

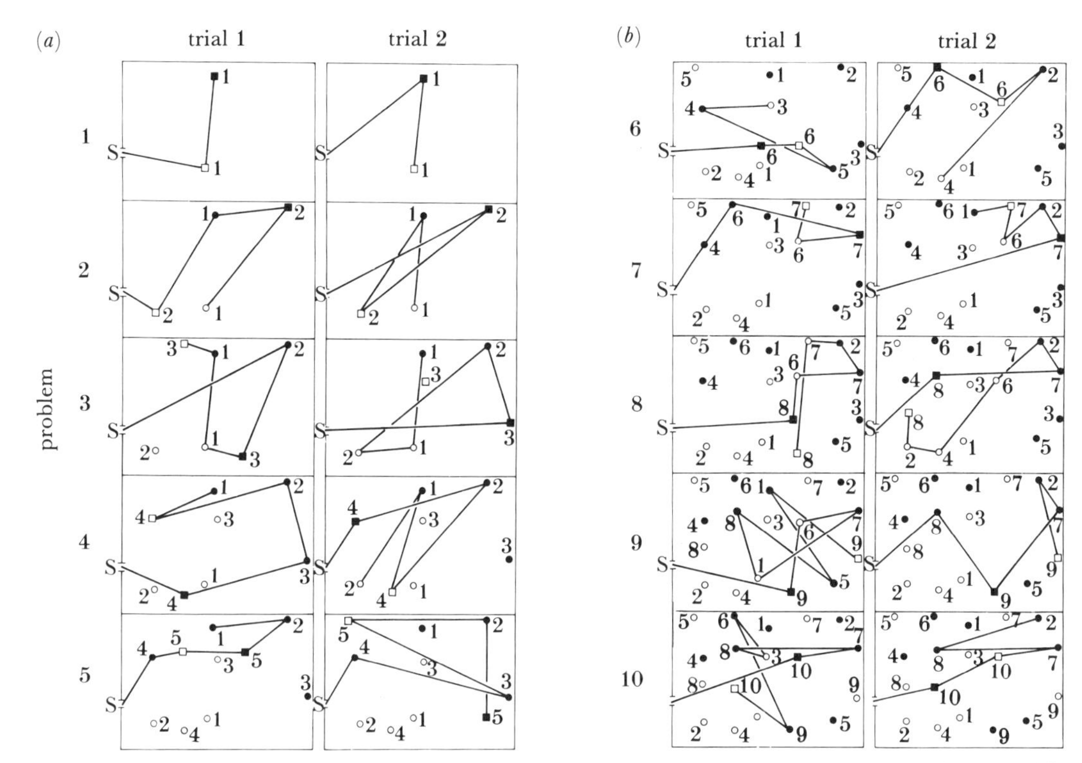

FIGURE 1. Maps of each trial in the first ten problems of test 1, excluding trial 3 of problem 7. The travel line shows the sequence in which the first five or more objects were first approached by any member of the group. Squares represent the newest objects; filled symbols the 'positive' objects; open symbols the 'negative' objects; and numbers the ordinal number of the problem on which each object was first introduced. Object locations varied from floor to ceiling, and the trees and tree limbs by which the monkeys traversed the room are not shown, so the actual relative travel-distances between objects are not very accurately portrayed. The locations of the two newest objects are incorrectly shown on trial 2 of problem 4; they should be the same as on trial 1 of that problem. The positive object was, however, approached first, as shown.

2 was one from which they had once obtained food; in 14 of these cases this object was one of the two newest positive objects. The most novel positive object was approached sooner than its equally novel, negative mate on trial 2 in 9 of the first 10 problems and in 13 of 16 problems overall; if each individual were to be scored separately on this same measure, their scores would average 75% 'correct' (range 70–80) in the first 10 problems, which is on a par with what other marmosets have achieved at the asymptote of their performance on Harlow's learning set task (Miles & Meyer 1956; Pournelle & Rumbaugh 1966). Four of the five individuals also spent more 15 s time intervals at any and all previously 'positive' objects than they spent at any and all negative objects, on every single trial (in the first 10 problems) on which none of the objects actually contained food. Still further, if one treats the various individuals as 'yoked controls' for one another and examines the correlation, within any given positive object, between the order in which each animal ate on trial 1 and the order in which it approached the same object on trial 2, the median Spearman ρ is 0.68 ($p < 0.001$). (In 12 of 16 problems whichever individual had been first to eat on trial 1 was among the first two animals to approach the positive object on trial 2, and in 13 of 16 problems this object was the first object in the room that this individual approached.) No comparable reliability in the social approach order was observed within the negative objects (median $\rho = 0.30$); and of all possible 240 correlations between trial 1 of any given positive object and trial 2 of the other positive objects the median

TABLE 1. THE BASIC DESIGN OF TEST 1, TOGETHER WITH DATA ON THE TOTAL NUMBER OF 15 S TIME INTERVALS SPENT, BY ALL FIVE GROUP MEMBERS, AT EACH OBJECT ON THOSE TRIALS IN WHICH NO OBJECTS CONTAINED FOOD

		pairs of objects																															
test		1		2		3		4		5		6		7		8		9		10		11		12		13		14		15		16	
day	prob.	A	B	A	B	A	B	A	B	A	B	A	B	A	B	A	B	A	B	A	B	A	B	A	B	A	B	A	B	A	B	A	B
2	1	6	5																														
4	2	4	2	67	7																												
6	3M	4	0	15	2	28	0																										
8	4	3	1	20	3	2	0	13	8																								
10	5M	5	6	4	2	4	0	16	1	25	10																						
12	6M	8	0	9	9	1	3	17	2	1	3	39	12																				
15	7	3	1	9	6	4	0	2	9	6	1	18	12	45	3																		
17	8M	0	0	3	7	1	0	4	12	23	2	0	1	8	2	26	8																
19	9	0	1	4	2	×	0	0	6	3	1	0	8	6	0	9	0	15	18														
21	10S	0	0	2	5		0	1	5	2	0	0	×	17	0	13	0	12	6	25	15												
24	11S	9*	0	1	8		0	2	9*	2	2	0		9	0	5	3	3	0	5	11	29	34										
26	12M	5	10*	11*	9		0	7	6	2	1	3		3	0	3	3	7	2	3	5	8	7	25	12								
28	13	7	3	1	3*		0	6*	0	9	0	0		6	0	4	12	2	0	6	5	8	6	6	4	42	17						
30	14M	7	0	0	0		7*	2	6	8	1	9*		3	0	7	0	4	1	2	6	8	10	4	0	16	0	34	7				
32	15	3	0	1	2		1	0	3	1	3	0		14*	0	1	0	8*	0	9	4	1	10	1	1	9	1	8	0	45	27		
34	16	1	3	2	1		2	0	2	6	0	0		7	5*	0	0	4	0*	6	4	2	4	3	1	9	0	3	11	12	8	15	6

M, new object moved ×, object removed S, new object swap location *, old object moved

Scores along the diagonal are the trial 3 data for problems 7 and 11, in which the 'positive' (A) object contained food on trial 2 as well as trial 1; otherwise, they are trial 2 data. The 'negative' (B) objects in problems 6–16 contained food odours on trial 1.

ρ is 0.25 and only 15% were as high as 0.68. Thus, individuals as well as the group as a whole showed clear evidence of one-trial learning, and the group data are not attributable to any fixed 'leadership hierarchy'.

In addition to discriminating between the 'positive' and 'negative' objects the animals also remembered the places in which they had found food. Thus, for example, the median ρ between the trial 1 order in which the animals ate and the trial 2 order in which they approached the location the positive object had occupied on trial 1 was 0.80 ($p < 0.001$), and this despite the fact that in some cases no test object was currently in that locus, and their approaches were sometimes made relatively late in a trial. In almost all problems, some food had spilled from the object on trial 1 as animals manipulated the object (twice, all of it spilled onto the ground, a metre or more beneath the object). The same individual that had retrieved it from any given spot stared at that same spot on trial 2, often making 'food calls' in the process and then looking at the spot the container had occupied on trial 1.

We found no evidence that very close spatial contiguity between food and cue object was essential for one-trial learning (cf. Jarvik 1956). It is clear, too, that the marmosets did not follow *either* a 'win–stay, lose–shift' strategy with respect to objects *or* the same sort of strategy with respect to locations; they followed *both* but simply attached a slightly lower priority to the latter. A major source of difficulty for them in more traditional tests, where two objects directly and repeatedly swap the same two locations is, we would think, not that they are 'naive' or 'poor learners' but that they learn too much, and therefore are confused by the strange 'rules of the game'. How could their feral counterparts manage to get back to a specific object in their native home range, which might exceed 30 ha in size, if they did not have a 'locational' strategy as well as an 'object' strategy? Indeed, we believe that the most accurate description of their achievements is that, with or without obvious reinforcement for so doing, they learn and remember the layout and structure of their environment: their 'rules of responding' being instantiations of that knowledge rather than substitutions for, or explanations of, it. Stated otherwise, learning to approach or avoid particular objects and locations is probably a special case of cognitive mapping rather than vice versa (Menzel 1978; Nadel & Willner 1980; Tolman 1948).

4. Test 2: learning with already-familiar objects

Here we sought to determine whether the results of the preceding tests were dependent upon the use of novel objects, or whether one-trial learning and 24 h memory could also be demonstrated with 'old' objects. The animals were given no cue on trial 1 (other than odour and other uncontrolled cues); the crucial question was what they would do on trial 2.

Method

On 16 occasions over the course of 20 trials, one object was selected from the 30 test objects that were left in the room from test 1 and baited with food on two successive trials. This object was selected at random, with the restriction that it not have contained food during the previous five trials. Odd-numbered objects ('problems') received their second baiting ('trial 2') at the same time that even-numbered ones were baited for the first time ('trial 1'). Throughout the first eight such problems, the locations of all objects remained constant. In the remaining eight problems, the location of the baited object was swapped with that of some other object on 'trial 2', this other object being selected by the same criteria as above. Between the eighth and ninth problems we gave the animals two runs in which no objects contained food or were moved. If the animals did not find the food within 15 min on trial 1 of any given problem we entered the test cage, tapped on the baited object as soon as one or more monkeys looked out of the home cage or otherwise attended to us, and then withdrew and sat down again at our observation stool about 1 m from the cage. (Usually they approached the object we had tapped without going to more than one other object first.)

Results and discussion

On 15 of the 16 critical test trials the marmosets approached whichever object contained food for the second time sooner than they approached the one that contained food for the first time; the latter objects were approached no sooner than chance. In 6 of the 8 problems on which the baited object was moved on trial 2, the same object was approached sooner than the other object which now occupied its old location; both exceptions resulted from the animals' apparent failure to detect the baited object for a time, for as soon as they spotted it they raced for it. The mean rank of the baited objects in the approach sequence on trial 2 was 5.7 (median = 2.5; 'chance' = 15.5; $p < 0.001$); only the two mentioned just above were approached later than expected by chance, and only one as late as it had been on trial 1. The objects that occupied their old (trial 1) locations were also approached sooner than expected by chance (mean rank = 6.3, median = 6; $p < 0.01$ difference from 15.5 and from their trial 1 score). Either the animals remembered the location in which they had found food, or they responded to the change in objects *per se* (see the next test), or both.

Almost all failures to go first to the objects that were baited for the second time involved prior response to the objects that had been baited on the immediately preceding problems; and within almost every trial there was a statistically significant correlation between the relative recency and frequency with which any given object had contained food and the order in which it was approached. Obviously, it is not necessary that an object be novel to demonstrate one-trial learning in *S. fuscicollis*; and although these animals have highly developed olfactory systems (Epple 1975), odour cues in the present study were not effective at long range.

5. Test 3: detection of changes in objects and their locations

Here we tested further how well the animals remembered both the visual appearances and the locations of test objects from one day to the next, as evidenced by their ability to detect change.

Method

A total of 15 trials were given, one per day. On five trials, two of the 30 test objects in the room were selected at random and their locations were directly swapped with one another; a coin toss decided which object was to contain food. On the remaining 10 trials one randomly picked object was removed from the room (for a single trial: it was repositioned in its old place the next time), and where it had been we placed a 'new' object that was visually identical to some third object that was already in the room. Either this 'new' object or its old visual counterpart contained food, five trials being given, in haphazard order, with each of these two possibilities. In this test and the preceding one all of the 30 objects were used at least once as a critical test object; thus, assuming that the animals could detect all of the above changes, it would be difficult to explain this fact without also assuming that they remembered all of the objects and all of their locations. (Note also that before test 2 only half of the objects had ever had food available in them, and only two had had food available in them more than once.)

Results and discussion

On the five trials in which the locations of two of the objects were directly swapped, the mean rank of the 10 objects in question, in the approach sequence, was 4.4 ($p < 0.001$). On the 10 trials in which the old object had been removed from the room and replaced by one that matched a third object somewhere else in the room, the mean rank for the former, 'new' object was 3.4 ($p < 0.001$) and on all 10 trials this object was hit before its 'old' counterpart.

The old counterpart objects were approached somewhat sooner than they had been on the previous trial (mean ranks = 15.1 and 19.8, respectively; $p = 0.05$) but no sooner than the average object (mean rank = 15.5). However, with more than half of them it was our strong impression that the monkeys had detected the similarity between them and the corresponding 'new' objects. For example, one animal seemed to eye the old object from a distance (too great to qualify as an unequivocal approach), went elsewhere for a while, repeated this sequence one or more times, and then not only it but several others went for the object en masse, making 'food calls' even before reaching it, and even though it did not in fact contain any food. It was as if they initially had only a vague premonition and thus hesitated in acting on it unless someone else also seemed to be looking at the same object.

6. Test 4: generalized delayed matching-to-sample and non-positional delayed response

In our next experiment, we tested these anecdotal impressions with a different procedure. Our primary purpose was, however, as follows. Given the animals' acute ability to detect whenever an object in a given locus was not the same object that was there on the preceding trial, any procedure that is aimed at experimentally disentangling 'object' discrimination from 'locational' discrimination by moving objects about from one trial to the next might create as many problems and confounds as it solves. We wished, therefore, to see whether we could demonstrate object discrimination without moving any of the test objects.

Method

Ten of the 30 test objects were discarded. Each day for 57 days a single trial was given; on each trial one of the 20 remaining test objects was selected at random (with the provision that it had not contained food for at least five trials), and baited. Before the test trial began, a different but visually almost identical ('sample') object was presented to the marmosets in their home cage. For the first 40 trials it contained five small pieces of candy and it was placed in the animal's cage and left there until they had found and eaten the candy. Following this it was removed and (about 1 min later) the door to the test room was opened. For the remaining trials we simply held the 'sample' object in front of the cage and let the monkeys watch as we dropped 10 pieces of candy into it. They could not touch the object and were not given any food. Then we stepped out of their home room and commenced the test run in the usual fashion. The sample objects were hidden from the animals' sight in all cases.

Results and discussion

The monkeys' relative latencies to approach the critical test object after having been given these cues ('day 0') are shown in figure 2, along with the corresponding scores for the same object on the immediately preceding and following trials. (Some of the last tests are omitted because we did not have complete information on trials following the single reinforced trial.) The mean rank approach score for day 0 was significantly lower than that of the average object

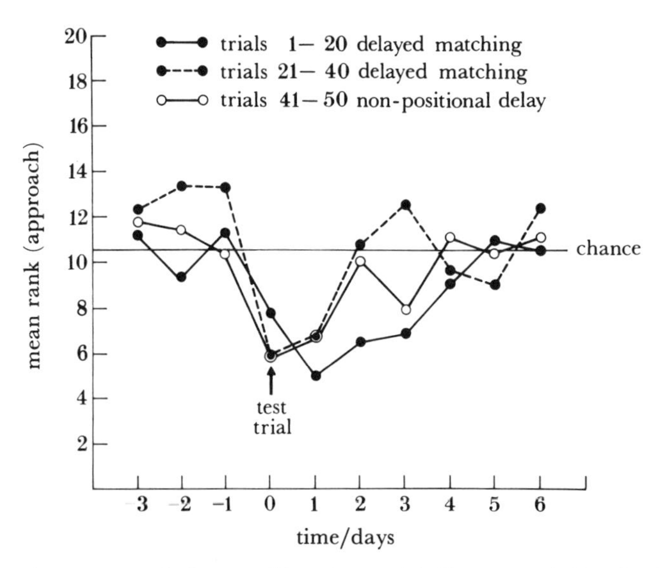

FIGURE 2. Relative latencies to approach the test objects in test 4, before, on and after the day they had encountered a 'sample' object in their home cage. (In this test there were 20 objects in the room; the first to be approached by any animal received a rank of 1, and so forth.)

and of the preceding trial even in the first 20 tests ($p < 0.05$ in all cases), and very clearly so for all 40 tests with the first (matching to sample) procedure ($p < 0.001$) and for the tests with the second (non-positional delayed response) procedure ($p < 0.001$). Overall, the critical test object was approached sooner on day 0 than on the preceding day on 85 % of the tests.

The data strongly suggest that the major 'improvement' that occurred with practice was attributable largely to the monkeys' decreasing tendency to go first to the objects that had been baited on the immediately preceding trials. In other words, memories of the previous day's run in the test room were initially a more important determinant of performance than memories of what had occurred a few minutes earlier in the home cage, and to reveal to an outside observer what we now presume they actually knew about the objects, they had to 'learn to forget'. The lack of any decrement in performance when we switched from the first procedure to the second shows, further, that food reinforcement during the cue-giving phase of the trial was not essential. In this experiment there was, coincidentally, little systematic relationship between the order in which the various individuals had obtained food from an object in their home cage and the order in which they approached the matching object in the test room (median $\rho = 0.25$).

7. Conclusions

Marmoset intelligence, as we see it, is whatever marmosets do, especially if it gives them an advantage over their competitors. The data strongly suggest that any other species might be hard put to do appreciably 'better' than *S. fuscicollis* on the same tasks, in the same test situation, with nothing more by way of prior instructions or training. It is worth noting further that in all of our experiments (including others in which we tested several social groups that contained infants) it has typically been the youngest animals that are most successful in getting the food. The rankings of marmosets relative to other species and of younger animals relative to older ones which one might obtain in more orthodox laboratory tasks might, of course, be very different; but it is our guess that such rankings might rest as much as anything on testing the animals alone in an isolated cubicle, with conditions that are strongly 'loaded' against any animal that is stressed or distracted by such isolation or that is not customarily a solitary forager.

Speaking more generally: although 'the ability to quickly solve new problems' or 'to apprehend the relationships between novel presented facts and organize one's actions accordingly' constitute the major ingredients of most definitions of intelligence, most 'standardized' tests of this hypothetical process, if not this definition as such, amount largely to roundabout definitions of *human* intelligence. Before animals are confronted with any given test situation, every species if not every individual has faced a different set of problems and developed a different set of strategies for solving them. These strategies might well all be optimal. Furthermore, depending largely upon how one's test situation and test criteria are loaded for or against it, almost any animal might in principle be made to rank either at the top or at the bottom of one's unidimensional quantitative scale of so-called correct responses. Such a point of view is directly complementary to the view that differences in intelligence are matters of degree rather than of kind: as, for example, is implied by Darwin's doctrine of 'mental continuity' and Harlow's analogous position that within any given species or age groups, types of learning that appear qualitatively distinct actually only represent different stages in the same, single, continuous quantitative function. It is, however, no less Darwinian a view; and in our opinion it provides a better 'explanation' of the present data. We do not, of course, imply that there are many investigators left today who would attempt to rank order species as to their relative 'general intelligence', let alone base such a ranking on any single index.

What is it about the behavioural organization of *S. fuscicollis* that makes the present sorts of tasks so simple for them? First of all, they are (even for primates) highly alert, visually curious and reactive to almost any sort of novelty or change. Secondly, they habituate to most objects quite rapidly, and show little or no spontaneous recovery of responsiveness on subsequent encounters with the same (unchanged) objects (Menzel & Menzel 1979, 1980). This is not a common propensity in Old World primates, at least in captivity; the 'advantage' that *S. fuscicollis* have here is that they are less manipulative. Thirdly, one class of objects towards which they do not thus habituate are those that proved to be edible, or to contain edibles, on last encounter (Menzel & Juno 1982). And, finally, being first in their group to locate food very likely provides additional (social) reinforcement, for foraging seems to be a team effort, involving a considerable amount of reciprocal altruism. All this, however, is an account of performance rather than of knowledge or perception. If the animals did not also know the nature and relative positions of the objects in question, these behavioural propensities would not do them much good. It is, of course, an open question how well they would do on problems that call for something other than a 'win–stay, lose–shift' strategy; and it is quite conceivable here that their skills are much more specialized, and less flexible, than those of (say) rhesus monkeys or chimpanzees.

It should not be forgotten, either, that assessments of an animal's intelligence rest not merely on the structure and apparent adaptiveness of any given performance, and how well it might fit some logical definition, but also on what we know (or assume) about how that performance originated. Regardless of what an insect can do (see, for example, Menzel & Erber 1978; Gould & Gould 1982), most people would hedge at calling it a cognitive genius, and the same is true of circus animals and robots. Partly for this reason, one of us (C.J.) is currently investigating the ontogenesis of one-trial learning in infant marmosets, within a 'normal' group context. It is a safe bet that some learning is involved; our guess is that much of it is social learning, particularly that which serves to wean infants from relying directly on their elders for food.

To most of humanity some animals quite simply look smarter and seem to act smarter than others. The fact that the smart-looking ones happen to be our own fairly close kin is no accident, and it makes us cautious about our own biases and prejudices. This, however, is only to say that judgments regarding intelligence might best be viewed as folk taxonomy rather than scientific taxonomy, and that no taxonomist should trust any external 'field marker' implicitly, let alone treat it as a sufficient definition of his or her genotypic concepts. Species do not have any platonic 'essence'; and neither does intelligence. At the same time, folk taxonomy is probably not completely mistaken, and some 'field markers' are indeed fairly trustworthy. In some cases they furnish as good an initial source of research ideas as any. Thus, for example, the first time we ever laid eyes on *S. fuscicollis* they impressed us as highly intelligent, largely because they seemed so alert, visually curious, and reactive towards us as strangers or indeed towards almost any novel event in their inferred visual and auditory field. All of the tests we devised amount, in retrospect, to little more than introducing some additional minor novelty or change into their environment, to test whether these first impressions were true or false, or in what senses they were true or false. Other researchers might have learned just as much about marmosets by some other strategy; but we doubt that we would have; and to say that *questions* about animal intelligence are 'outdated' or 'scientifically useless' would be short-sighted.

References

Dawson, G. A. 1979 The use of time and space by the Panamanian tamarin (*Saguinus oedipus*). *Folia primat.* **31**, 253–284.

Epple, G. 1975 The behavior of marmoset monkeys (Callitrichidae). In *Primate behavior* (ed. L. A. Rosenblum), vol. 4, pp. 195–239. New York: Academic Press.

Ford, S. M. 1980 Callitrichids as dwarfs and the place of the Callitrichidae in Platyrrhini. *Primates* **21**, 31–43.

Garber, P. A. 1980 Locomotor and feeding ecology of the Panamanian tamarin (*Saguinus oedipus geoffroyi*, Callitrichidae, Primates). *Int. J. Primat.* **1**, 185–201.

Gould, J. L. & Gould, C. G. 1982 The insect mind: physics or metaphysics? In *Animal mind – human mind* (ed. D. R. Griffin). Berlin: Dahlem Konferenzen, Springer-Verlag.

Harlow, H. F. 1949 The formation of learning sets. *Psychol. Rev.* **56**, 51–65.

Izawa, K. 1978 A field study of the black-mantle tamarin (*Saguinus nigricollis*). *Primates* **19**, 241–274.

Jarvik, M. E. 1956 Simple color discrimination in chimpanzees: effect of varying contiguity between cue and incentive. *J. comp. Physiol. Psychol.* **49**, 492–495.

Johnston, T. D. 1981 Contrasting approaches to a theory of learning. *Behav. Brain. Sci.* **4**, 125–139.

Kamil, A. C. & Sargent, T. (ed.) 1981 *Foraging behavior: ecological, ethological and psychological approaches.* New York: Garland Press.

Kamil, A. C. & Yoerg, S. I. 1982 Learning and foraging behavior. In *Perspectives in ethology* (ed. P. P. G. Bateson & P. H. Klopfer), vol. 5, pp. 325–364. New York: Plenum.

Levine, M. 1959 A model of hypothesis behavior in discrimination learning set. *Psychol. Rev.* **66**, 353–366.
Menzel, C. R. & Menzel, E. W. 1980 Head-cocking and visual exploration in marmosets, Saguinus fuscicollis. *Behaviour* **75**, 219–234.
Menzel, E. W. 1971 Group behavior in young chimpanzees: Responsiveness to cumulative novel changes in a large outdoor enclosure. *J. comp. Physiol. Psychol.* **74**, 46–51.
Menzel, E. W. 1978 Cognitive mapping in chimpanzees. In *Cognitive processes in animal behavior* (ed. S. H. Hulse, H. Fowler & W. K. Honig), pp. 375–422. Hillsdale, N.J.: Lawrence Erlbaum Associates, Inc.
Menzel, E. W. & Juno, C. 1982 Marmosets (Saguinus fuscicollis): Are learning sets learned? *Science, Wash.* **217**, 750–752.
Menzel, E. W. & Juno, C. 1984 Are learning sets learned? Or: perhaps no nature–nurture issue has any simple answer. *Anim. Learn. Behav.* **12**, 113–115.
Menzel, E. W. & Menzel, C. R. 1979 Cognitive, developmental and social aspects of responsiveness to novel objects in a family group of marmosets (*Saguinus fuscicollis*). *Behaviour* **70**, 251–278.
Menzel, R. & Erber, J. 1978 Learning and memory in bees. *Scient. Am.* **239**, 80–87.
Miles, R. C. & Meyer, D. R. 1956 Learning sets in marmosets. *J. comp. Physiol. Psychol.* **49**, 219–222.
Milton, K. 1981 Distribution patterns of tropical plant foods as an evolutionary stimulus to primate mental development. *Am. Anthrop.* **83**, 534–548.
Mittermeier, R. A. & Roosmalen, M. G. M. van 1981 Preliminary observations on habitat utilization in eight Surinam monkeys. *Folia primat.* **36**, 1–39.
Moynihan, M. 1976 *The New World primates*. Princeton, N.J.: Princeton University Press.
Nadel, L. & Willner, J. 1980 Context and conditioning: A place for space. *Physiol. Psychol.* **8**, 218–228.
Olton, D. 1979 Mazes, maps and memory. *Am. Psychol.* **34**, 583–596.
Pournelle, M. B. & Rumbaugh, D. M. 1966 A comparative study of learning in the squirrel monkey, the golden marmoset and the cotton-topped tamarin. *Am. Psychol.* **21**, 901.
Richard, A. F. 1981 Changing assumptions in primate ecology. *Am. Anthrop.* **83**, 517–532.
Schrier, A. M. & Thompson, C. R. 1984 Are learning sets learned? A reply. *Anim. Learn. Behav.* **12**, 109–112.
Szalay, F. S. & Delson, E. 1979 *Evolutionary history of the primates*. New York: Academic Press.
Terborgh, J. W. 1983 *Five New World primates: A study in comparative ecology*. Princeton, N.J.: Princeton University Press.
Tolman, E. C. 1948 Cognitive maps in rats and men. *Psychol. Rev.* **55**, 189–208.

Discussion

P. Garrud (*University of St Andrews, U.K.*). I should like to ask about something puzzling in the results from the one-trial object discrimination task. Why do the marmosets return sooner to the previously baited member of any particular pair than to the unbaited object, not only on the trial immediately following the baited trial but also on nearly all the ensuing trials? This seems surprising when:

(i) the animals ate all the food on the first trial in any case;

(ii) the animals have revisited that object at least once (on the next trial) and found it empty;

(iii) they are presented with a series of problems, all of which have this general form: that the new object baited on any trial is never baited thereafter.

E. W. Menzel. This might be a redescription of the results that you note rather than an explanation, but I'd say that even without special training on our part *Saguinus fuscicollis* have very strong 'win–stay' strategies and excellent memories. The results might be surprising or puzzling in the light of what other animals do in other situations, but they are quite consistent with Terborgh's field data on the foraging habits of wild *S. fuscicollis* and with what our own laboratory animals do every day in their home cages and in our other test conditions. They become less puzzling, too, if one thinks of some other well-known analogies. For example, rats and hummingbirds tend to 'win–shift' or 'alternate' with respect to locations at the outset of many experiments, and to persist at this to some degree and for some time despite an experimenter's not rewarding them for it with food. Their specific response strategy is different, but in other respects there are many commonalities.

Actually, too, the effects of a single reinforced trial were not as long-lasting as you suggest: after several unreinforced trials with the same objects these effects were no longer statistically significant; and with repeated tests of the same sort they diminished. (See here figure 2 for a graphic representation of these points, from our fourth experiment.) Although it initially goes against their grain, marmosets can also learn to 'win–shift', just as rats and hummingbirds can learn to 'win–stay'. Thus, for example, in one of our later experiments our animals learned to go sooner on trial 2 to objects that had had a piece of inedible paper in them on trial 1 (and that were baited with food on trial 2) than to objects that had contained food on trial 1 (but contained nothing on trial 2).

Phil. Trans. R. Soc. Lond. B **308**, 159–176 (1985) [159]
Printed in Great Britain

Signs of intelligence in cross-fostered chimpanzees

By Beatrix T. Gardner and R. A. Gardner
Department of Psychology, University of Nevada, Reno, Nevada 89557, *U.S.A.*

In cross-fostering, the young of one species are reared by adults of another, as in the classical ethological studies of imprinting and song-learning. In our laboratory, infant chimpanzees were reared under human conditions that included two-way communication in American Sign Language (A.S.L.), the gestural language of the deaf in North America. A large body of evidence from five chimpanzees demonstrated stage by stage replication of basic aspects of the acquisition of speech and signs by hearing and deaf children. Here we review evidence that, under double-blind conditions: (i) the chimpanzees communicated information in A.S.L. to human observers; (ii) independent human observers agreed in their identification of the chimpanzee signs, (iii) the chimpanzees could use the signs to refer to natural language categories: DOG for any dog, FLOWER for any flower, SHOE for any shoe.

On 21 June, 1966, an infant chimpanzee arrived in our laboratory. We named her Washoe, for Washoe County, the home of the University of Nevada. To a casual observer Washoe's new home may not have looked very much like a laboratory. In fact, it was the Gardner residence in the suburbs of Reno, purchased as a faculty home some years earlier, a small, one storey, brick and wood home with an attached garage and a largish back yard. To that same casual observer, Washoe's daily life may not have looked much like laboratory routine, either. It was rather more like the daily life of human children of her age in the same suburban neighbourhood.

Washoe was about ten months old when she arrived in Reno, and almost as helpless as a human child of the same age. In the next few years she learned to drink from a cup and eat at a table with forks and spoons. She also learned to set and clear the table and even to wash the dishes, in a childish way. She learned to dress and undress herself and she learned to use the toilet to the point where she seemed embarrassed when she could not find a toilet on an outing in the woods, eventually using a discarded coffee pot that she found on a hike. She had the usual children's toys and was particularly fond of dolls, kissing them, feeding them, and even bathing them. She was interested in picture-books and magazines almost from the first day and she would look through them by herself or with a friend who would name and explain the pictures and tell stories about them. The objects and activities that most attracted her were those that most engaged the grown-ups. She seemed to be fascinated by household tools, eventually acquiring a creditable level of skill with hammers and screwdrivers.

Washoe lived in a second-hand housetrailer, parked in a garden behind the house. With a few minor alterations, it was the same trailer that its previous owners had used as a travelling home. It had the same living-room and bedroom furniture and the same kitchen and toilet facilities. Someone came in to the trailer to check her each night and all through the night, every night, someone listened to her by means of an intercom connected to the Gardner home.

To a casual observer the greatest departure from the world of most human children would probably have been the silence. Modern man is a noisy member of the animal kingdom. Old or young, male or female, wherever you find two or more human beings they are usually vocalizing. By contrast, chimpanzees are usually silent. They seldom vocalize unless they are excited (Yerkes & Yerkes 1929, pp. 301–309; Goodall 1965). Washoe was also very silent and so were her human companions. The only language that we used in her presence was American Sign Language (A.S.L.). There were occasional lapses, as when outside workmen or her paediatrician entered the laboratory, but the lapses were brief and rare.

When Washoe was present, all business, all casual conversation was in A.S.L. Everyone in Washoe's foster family had to be fluent enough to make themselves understood under the sometimes hectic conditions of life with this lively youngster. Visits from non-signers were strictly limited. Some university professors declined to enter the laboratory when they realized that speaking was against the rules. Others discovered within minutes that the discipline was too much for them, 'I know I musn't talk, but...' and their visits were shortened accordingly. A few truly social beings invented their own sign language on the spot. Eventually, we hit on the procedure of teaching visitors a few key phrases that they could repeat indefinitely, I VERY HAPPY MEET YOU, YOU PRETTY GIRL, WASHOE SMART GIRL, and so forth. Visitors from the deaf community who were fluent in A.S.L. were always a welcome relief.

The rule of sign-language-only required some of the isolation of a field expedition. We lived and worked with Washoe on that corner of suburban Reno as if at a lonely outpost in a hostile country. We were always avoiding people who might speak to Washoe. On outings in the country, we were as stealthy and cautious as Indian scouts. On drives in town, we wove through traffic like undercover agents. We could stop at a Dairy Queen or a MacDonald's fast-food restaurant, but only if they had a secluded car park in the back. Then one of Washoe's companions could buy the treats while another waited with her in the car. If Washoe was spotted, the car drove off to return later for the missing passenger and the treats, when the coast was clear.

1. Cross-fostering

The first scientist to comment on the Reno laboratory was Winthrop Kellogg, the great pioneer in this field.

> 'Apes as household pets are not uncommon and several books by lay authors attest to the problems involved.... But pet behavior is not child behavior, and pet treatment is not child treatment. It is quite another story, therefore, for trained and qualified psychobiologists to observe and measure the reactions of a home-raised pongid amid controlled experimental home surroundings. Such research is difficult, confining, and time-consuming. ... Although often misunderstood, the scientific rationale for rearing an anthropoid ape in a human household is to find out just how far the ape can go in absorbing the civilizing influences of the environment. To what degree is it capable of responding like a child and to what degree will genetic factors limit its development?' (Kellogg 1968, p. 423).

Earlier, the Kelloggs had outlined the requirements for cross-fostering.

> 'One important consideration upon which we would insist was that the psychological as well as the physical features of the environment be entirely of a human character. That

is, the reactions of all those who came in contact with the subject, and the resulting stimulation which these reactions afforded the subject, should be without exception just such as a normal child might receive. Instances of anthropoid apes which have lived in human households are of course by no means unknown. But in all the cases of which we have any knowledge the 'human' treatment accorded the animals was definitely limited by the attitude of the owner and by the degree of his willingness to be put to boundless labor. It is not unreasonable to suppose, if an organism of this kind is kept in a cage for a part of each day or night, if it is led about by means of a collar and a chain, or if it is fed from a plate upon the floor, that these things must surely develop responses which are different from those of a human. A child itself, if similarly treated, would most certainly acquire some genuinely unchildlike reactions. Again, if the organism is talked to and called like a dog or a cat, if it is consistently petted or scratched behind the ears as these animals are often treated, or if in other ways it is given pet stimuli instead of child stimuli, the resulting behavior may be expected to show the effects of such stimulation.

'In this connection it was our earnest purpose to make the training of the ape what might be called incidental as opposed to systematic or controlled training. What it got from its surroundings it was to pick up by itself just as a growing child acquires new modes of behavior. We wished to avoid deliberately teaching the animal, trial by trial, a series of tricks or stunts which it might go through upon signal or command. The things that it learned were to be its own reactions to the stimuli about it. They were furthermore to be specifically responses to the household situation and not trained-in or meaningless rituals elicited by a sign from a keeper. The spoon-eating training, to take a concrete example, was to be taken up only in a gradual and irregular manner at mealtime, as the subject's muscular coordination fitted it for this sort of manipulation. We would make no attempt to set it down at specified intervals and labor mechanically through a stated number of trials, rewarding or punishing the animal as it might succeed or fail. Such a proposed procedure, it will be readily seen, is loose and uncontrolled in that it precludes the obtaining of quantitative data on the number of trials necessary to learn, the number of errors made, or the elapsed time per trial. It has the advantage, nevertheless, of being the same sort of training to which the human infant is customarily subjected in the normal course of its upbringing' (Kellogg & Kellogg 1933, pp. 12–13).

It seems as if no form of behaviour is so fundamental or so distinctively species-specific that it is not deeply sensitive to the effects of early experience. Ducklings, goslings, lambs, and many other young animals learn to follow the first moving object that they see, whether it is their own mother, a female of another species, or a shoebox. The mating calls of many birds are so species-specific that an ornithologist can identify them by their calls alone without seeing a single feather. Distinctive and species-specific as these calls may be, they, too, depend upon early experience.

Niko Tinbergen and his students have made the British Herring Gull one of the most thoroughly studied of all animals. Normally reared Herring Gulls spend the entire year in Britain; they never migrate south for the winter. The lesser Black-backed Gull also breeds on British shores, but members of their species normally migrate south to the sea-coasts of Spain, Portugal, and northwest Africa. Harris (1970) arranged for experimental cross-fostering by placing Herring Gull eggs in the nests of lesser Black-backed Gulls and *vice versa*, banding the chicks after hatching so that they could be identified as adults. Many cross-fostered Herring Gulls were recovered on the coasts of Spain and Portugal, even though the effects of cross-fostering failed to induce many lesser Black-backed Gulls to overwinter in Britain.

How about our species, how much does our common humanity depend upon our common experience of a species-typical human childhood? The question is so tantalizing that even alleged but unverified cases of human cross-fostering, such as Itard's account of Victor, 'the wild boy of Aveyron', attract serious scholarly attention. Many, presumably insurmountable, ethical and practical difficulties stand in the way of experimentally controlled, or even verifiable cases of human children cross-fostered by non-human beings. The Kelloggs were the first to attempt the logical alternative; a form of cross-fostering in which the subjects are chimpanzees and the foster parents are human beings.

2. Introduction of A.S.L.

Before Project Washoe, there were, altogether, four professional research projects that had followed Kellogg's procedures of cross-fostering (see Kellogg 1968). In all four cases, the infant chimpanzees thrived, and all aspects of their behavioural development resembled the development of human children, with one striking exception. There was hardly any development of spoken language. In the most successful case, the chimpanzee Viki spoke only four words, MAMA, PAPA, CUP and UP, after nearly seven years of intensive exposure to English together with additional sophisticated, thorough, and ingenious teaching (Hayes & Nissen 1971).

For decades, the failures of Gua and Viki to learn to speak were cited and recited to support the traditional doctrine of absolute, unbridgeable discontinuity between human and non-human. Other scientists, aware of the silent habits of chimpanzees, looked for a technique that would not require speech. This was the innovation of Project Washoe. For the first time, the foster family used a gestural rather than a vocal language.

With the introduction of A.S.L., the line of research pioneered by the Kelloggs and the Hayeses moved forward, dramatically. In 51 months, Washoe acquired at least 132 signs of A.S.L. and used them for classes of referents rather than specific exemplars. Thus, DOG was used to refer to live dogs and pictures of dogs of many breeds, sizes, and colours, as well as the sound of barking by an unseen dog. OPEN was used to ask for the opening of doors to houses, rooms, cupboards, or the lids of jars, boxes, bottles, and even (an invention of Washoe's) for turning on a water tap. Washoe also understood many more signs than she used herself (Gardner & Gardner 1975).

She signed to friends and to strangers. She signed to herself and to dogs, cats, toys, tools, even to the trees. She asked for goods and services, and she also asked questions about the world of objects and events around her. When Washoe had about eight signs in her expressive vocabulary, she began to combine them into meaningful phrases. YOU ME HIDE, and YOU ME GO OUT HURRY were common. She called her doll, BABY MINE; the sound of a barking dog, LISTEN DOG; the refrigerator, OPEN EAT DRINK; and her potty-chair, DIRTY GOOD. Along with her skill with cups and spoons, and pencils and crayons, her signing developed stage for stage much like the speaking and signing of human children (Gardner & Gardner 1971, 1974a; Van Cantfort & Rimpau 1982).

Project Washoe was followed by a second, more advanced, venture in cross-fostering

(Gardner & Gardner 1978, 1980). Washoe herself was captured wild in Africa and arrived in Reno when she was about 10 months old. Moja, Pili, Tatu, and Dar, were born in American laboratories and each arrived within a few days of birth. In general, the human participants in the second project had a higher level of expertise in A.S.L. and chimpanzee psychology because some of them were veterans of Project Washoe. Many of the new recruits were deaf or the hearing offspring of deaf parents or had other extensive experience with A.S.L. before they joined the staff. All had learned A.S.L. and studied the procedures and results of Project Washoe beforehand. And, while Washoe was the only chimpanzee in Reno, the foster chimpanzees of the second project had each other as frequent companions. The chimpanzees used the signs of A.S.L. with each other and they learned new signs from each other. At a later stage in the Fouts laboratory, the infant Loulis acquired, and used appropriately, at least 47 signs that he could only have learned from Washoe, Moja, Tatu, or Dar (Fouts *et al.* 1984).

Most commentators have acknowledged that this line of research has demonstrated a significant degree of intellectual continuity between cross-fostered chimpanzees and human children (cf. Bronowski & Bellugi 1970; Brown 1970; Dingwall 1979; Donahoe & Wessells 1980; Griffin 1976; Hewes 1973; Hill 1978; Hockett 1978; Kellogg 1969; Lieberman 1984; Marler 1969; Stokoe 1978; Thorpe 1972; Van Cantfort & Rimpau 1982; Watt 1974). Even among those who remain faithful to the doctrine of ultimate cognitive discontinuity, most have conceded that there is now more evidence of continuity than they would have expected.

Teaching

The development of verbal behaviour, as we know it in the human case, is inextricably bound up in the rest of the conditions of a human childhood. We doubt whether anything comparable could develop under other conditions. It seems equally unlikely that the other aspects of human intellectual growth could flourish without the development of verbal behaviour. Thus, the purpose of A.S.L. in the young lives of Washoe, Moja, Pili, Tatu, and Dar, was to satisfy the requirements of cross-fostering. It was a means, rather than an end in itself. Without two-way communication in a naturally occurring human language, Kellogg's objectives could not be met. The conditions of enhancement would have to fall far short of a human childhood. At the same time, almost any measurement of the intellectual development of human children must soon include verbal behaviour; comparisons between the young chimpanzees and the young children had to include verbal behaviour.

By contrast, much of the other research that followed in the wake of Project Washoe (for example, Premack 1971; Rumbaugh *et al.* 1973; Terrace *et al.* 1979), has assumed discontinuity between verbal behaviour and the rest of intelligent behaviour. By and large, these investigators have sought to separate language from the rest of behaviour by dint of rigorous theoretical analysis and to train their chimpanzee subjects to perform certain narrowly defined tasks based on *a priori* definitions of language. No comparisons have been made with human children performing the same tasks under the same conditions, indeed, no such comparisons are possible.

At the height of the, so called. 'Chomskian revolution' in psycholinguistics, it was frequently claimed that human children acquire their first language with incredible speed by the innate unfolding of a uniquely human mental process, and more or less independently of adult input (Lenneberg 1967, p. 137; McNeill 1966; Moore 1973, p. 4). Needless to say, this claim was always in conflict with common experience. More recently, a large body of painstaking research has supported the more traditional view, that human parents teach their children. As Snow (1977) puts it

> 'the first descriptions of mothers' speech to young children were undertaken in the late sixties in order to refute the prevailing view that language acquisition was largely innate and occurred almost independently of the language environment. The results of those mothers' speech studies may have contributed to the widespread abandonment of this hypothesis about language acquisition, but a general shift from syntactic to semantic–cognitive aspects of language acquisition would probably have caused it to lose its central place as a tenet of research in any case. [p. 31] ...all language-learning children have access to this simplified speech register. No one has to learn to talk from a confused, error-ridden garble of opaque structure. Many of the characteristics of mothers' speech have been seen as ways of making grammatical structure transparent, and others have been seen as attention-getters and probes to the effectiveness of the communication' [p. 38].

In teaching sign language to Washoe, Moja, Pili, Tatu, and Dar we observed human parents with young children and we imitated them. There was constant chatter about the everday events and objects that might interest the young chimpanzees. Many of the comments were aimed at teaching vocabulary, for example, THAT CHAIR, SEE PRETTY BIRD, MY HAT. Many events and objects were introduced, just so we could sign about them. There were frequent questions to see what was getting across, and we tried hard to answer all the youngsters' questions. By expanding on fragmentary utterances we could use the fragments to teach and to probe. We also followed the parents of deaf children in using an especially simple and repetitious register of A.S.L. and making signs on the youngsters' bodies to capture their attention (cf. Schlesinger & Meadow 1972).

Recording

While teaching was spontaneous and informal, as in a human nursery, the methods of recording results were precisely defined and meticulously followed. Each sign had to meet detailed criteria of form and usage before it was listed as a reliable item of vocabulary. In terms of form it had to correspond to a sign made by human adults, or to an immature variant. Decisions were guided by the judgement of fluent signers who were also familiar with signing in young children. In the case of usage, spontaneity was defined in terms of informative prompting. If the sign was produced by the chimpanzee subject without informative prompting, such as direct modelling or guidance that would induce any portion of the target sign, then it was judged to be spontaneous. To be appropriate, however, it had to be prompted by the verbal and situational context, and the presence of a suitable addressee (Gardner & Gardner 1971, 1975, 1978, 1980).

Appropriate usage was judged on the basis of context notes, with the understanding that infant usage can be either more narrow or more broad than adult usage. Nevertheless, the chimpanzee usage had to have some major overlap with adult usage. Thus, Washoe and Tatu used OUT both for leaving and for entering their quarters in these early records (it was only later that they divided this referential domain between OUT and IN). To refer to signs here and elsewhere throughout our writing, we have used as a gloss, the nearest English equivalent to the sign. This is always a single English word, and it is the gloss listed for that particular sign in one of three references (Fant 1977; Stokoe *et al.* 1976; Watson 1973).

3. Structure and function

Until recently, studies of the development of verbal behaviour in human children concentrated on the acquisition of grammar to the near exclusion of vocabulary. The bulk of the evidence has been based on naturalistic methods such as diary records, inventories of phrases, and samples of utterances. The same naturalistic methods were used with Washoe, Moja, Pili, Tatu, and Dar and comparisons can be found in Gardner & Gardner (1969, 1971, 1974*a*, 1975, 1978, 1980) and in work still in progress. With respect to structure, when Van Cantfort & Rimpau (1982) reviewed the evidence in detail they found that, when the same rules of evidence are applied to both sets of data, the same results are found for the early utterances of children and chimpanzees. As Van Cantfort & Rimpau also showed, arguments for discontinuity that seem to be based on the same data have, instead, been based on a double-standard of evidence. 'Too often the comparisons that are cited are between three-year-old chimpanzees and university-level human beings or between observations of chimpanzee utterances and idealized, theoretical conceptions of human linguistic competence' (p. 65).

Although concern with grammar has occupied so much of the efforts of developmental psycholinguists, in our view it would be a mistake for psychobiologists to neglect method and theory in the study of reference. For, if the development of human verbal behaviour requires any significant expenditure of biological resources, then it must confer selective advantages on its possessors. To confer any selective advantage, however, a biological trait must operate on the world in some way; it must be instrumental in obtaining benefit or avoiding harm. If clarifying one's ideas confers selective advantage it must be because, in some way, clarified ideas provide superior means for operating in the biological world. As for establishing social relations, a system of displays and cries is sufficient to maintain group cohesiveness in most animals. The selective advantage of a wider variety of signals would seem to be the communication of more information. But, unless verbal behaviour refers to objects and events in the external world, it cannot communicate information and it cannot have any such selective advantage. From this point of view, reference is the Darwinian function of verbal behaviour, and the function of grammar or structure in verbal behaviour must be to enlarge the scope and increase the precision of reference.

4. A test of communication†

When Washoe was 27 months old she made a hole in the then flimsy inner wall of her house trailer. The hole was located high up in the wall at the foot in her bed. Before we repaired the hole she managed to lose a toy in the hollow space between the inner and outer walls. When Allen Gardner arrived that evening she attracted his attention to an area of the wall down below the hole at the level of her bed, signing OPEN OPEN many times over that area. It was not hard for Allen Gardner to understand what the trouble was and eventually to fish out the toy. When the toy was found, it was exciting to realize that a chimpanzee has used a human

† See Gardner & Gardner (1984) for a complete description of the procedure and results of these tests.

language to communicate truly new information. It was not long before such situations became commonplace. For example, Washoe's playground was in the garden behind a single-storey house. High in her favourite tree, Washoe was often the first to know who had arrived at the front of the house and her companions on the ground learned to rely on her to tell them who was arriving and departing.

Washoe could tell her human companions things that they did not already know. This is what Clever Hans could not do. Clever Hans, it will be remembered, was a German horse that seemed to do arithmetic by tapping out numbers with his hoof. Not the circus trainers nor the cavalry officers, not the veterinarians nor the zoo directors, not even the philosophers and the linguists who studied the case could explain how Clever Hans did it. Nevertheless an experimental psychologist, Oskar Pfungst (1911), unravelled the problem with the following test. Pfungst whispered one number into Clever Hans left ear and Herr von Ost, the trainer, whispered a second number into the horse's right ear. When Clever Hans was the only one who knew the answer, he could not tap out the correct sums. He could not tell his human companions anything that they did not already know.

Since then, controls for 'Clever Hans errors' have been standard procedure in comparative psychology†. To date most, if not all, research on human children has been carried out without any such controls. It is as if students of child development believed that, whereas horses and chimpanzees may be sensitive to subtle non-verbal communication, it is safe to assume that human children are totally unaffected.

Method

Early in Project Washoe we devised vocabulary tests to demonstrate that chimpanzees could use the signs of A.S.L. to communicate information. The earliest versions are described in Gardner & Gardner (1971), pp. 158–161, 1974*a*, pp. 11–15, 1974*b*, pp. 160–161). The first objective of these tests was to demonstrate that the chimpanzee subjects could communicate information under conditions in which the only source of information available to a human observer was the signing of the chimpanzees. To accomplish this, nameable objects were photographed on 35 mm slides. During testing, the slides were back-projected on a screen that could be seen by the chimpanzee subject, but could not be seen by the observer. The slides were projected in a random order that was changed from test to test so that the order could not be memorized either by the observer or by the subject.

The second objective of these tests was to demonstrate that independent observers agreed with each other. To accomplish this, there were two observers. The first observer (O_1) served as interlocutor in the testing room with the chimpanzee subject. The second observer (O_2), was stationed in a second room and observed the subject from behind one-way glass, but could not see the projection screen. The two observers gave independent readings; they could not see each other and they could not compare observations until after a test was completed.

The third objective of these tests was to demonstrate that the chimpanzees used the signs to refer to natural language categories: that the sign DOG could refer to any dog, FLOWER to any flower, SHOE to any shoe, and so on (Rosch & Lloyd 1978; Saltz *et al.* 1972, 1977). This was accomplished by preparing a large library of slides to serve as exemplars. Some of the slides

† In modern times a striking exception to this rule has been the work of Terrace (1979) with the chimpanzee Nim, which included no controls whatever for 'Clever Hans errors'.

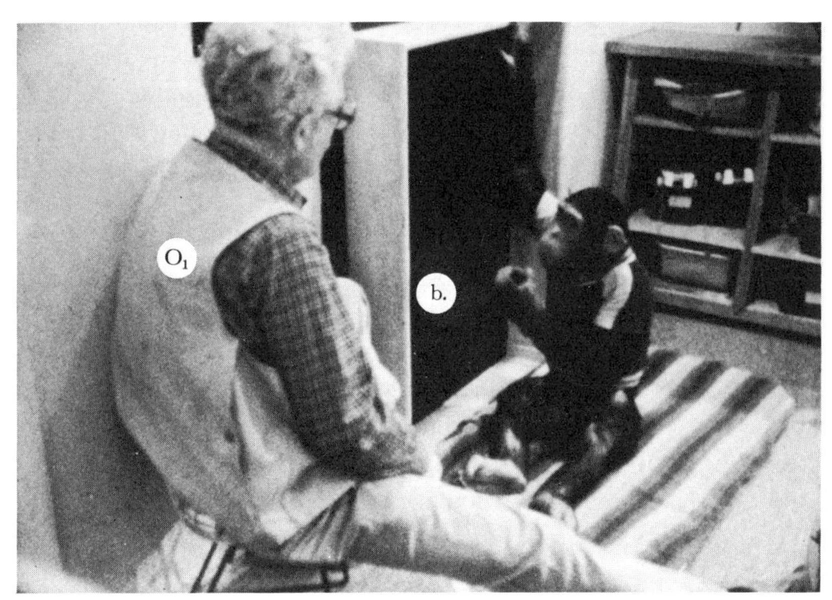

FIGURE 1. Room 1 of the vocabulary testing apparatus used with Moja, Tatu, and Dar. Chimpanzee Dar sits in front of the projection screen, which is recessed within the cabinet. O_1, seated beside the cabinet, can see Dar's signs but cannot see the projecton screen. By pressing the white push-button (b.), Dar makes a picture appear on the screen. (The vocabularly testing apparatus used with Washoe was slightly different; see Gardner & Gardner, 1974*a*, pp. 11–16).

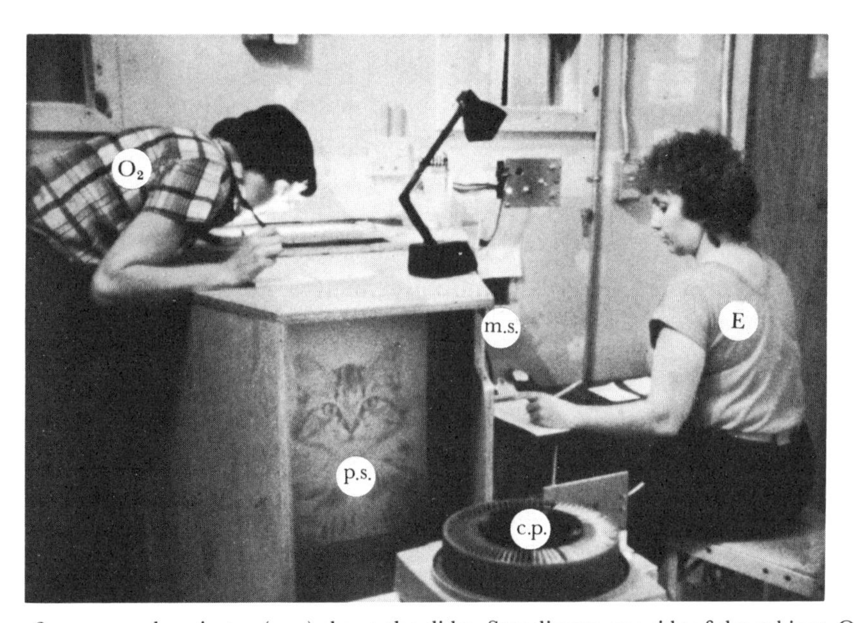

FIGURE 2. Room 2: a carousel projector (c.p.) shows the slides. Standing to one side of the cabinet, O_2 can see the subject through the one-way glass window but cannot see the projection screen (p.s.). O_2 writes down what the subject signs and passes the message slip to the experimenter (E), who also receives written messages from O_1, via the message slot (m.s.). After receiving a message slip from O_1 and O_2, E. steps the carousel to the slide for the next trial.

were used in pre-tests that served to adapt subjects, observers, and experimenters to the testing procedure. The slides that were reserved for the tests were never shown during pre-tests so that the first time that a particular chimpanzee subject saw any one of the test slides was on a test trial and no test slide was shown on more than one test trial. Consequently, there was no way that a subject could get a correct score by memorizing particular pairs of exemplars and signs. That is to say, scores on these tests depended upon the ability to name new exemplars of natural language categories.

The vocabulary items that appeared in Moja's single test and in the second test of each of the other three subjects are listed in table 1. All of the vocabulary items tested in this way had to be names of picturable objects (see Gardner & Gardner 1984). Differences among the subjects in table 1 reflect differences in their vocabularies as well as a strategy of overlapping tests that would sample the range of picturable objects in the vocabularies without making the tests excessively long. For each test we chose four exemplars of each vocabulary item to illustrate the range of objects that a subject could name with the same sign. Different breeds represented CAT and DOG, different species repesented BIRD and BUG, different makes and models represented CAR. The number of vocabularly items and the resulting number of trials (items × exemplars) appear in table 2.

The correct sign for each vocabulary item was designated in advance of the tests. That sign and that sign only was scored as correct for that item. Although there were aspects of the pictures for which superordinate terms, such as FOOD, or descriptive terms such as BLACK, might be scored, neither the presence or absence nor the correct or incorrect use of such terms was considered in the scoring of these tests.

Most of the replies consisted of a single sign which was the name of an object. Sometimes, the single noun in the reply was contained in a descriptive phrase, as when Tatu signed RED BERRY for a picture of cherries, or when Dar signed THAT BIRD for a picture of a duck. These replies contained only one object name and that was the sign that was scored as correct or incorrect. Occasionally, there was more than one object name in a reply, as when Washoe signed FLOWER TREE LEAF FLOWER for a picture of a bunch of daisies. In such cases, the observers designated a single sign for scoring (usually the first) without looking at the picture themselves. For each trial and each observer, then, one sign and one sign only, in each report was used to score agreement between O_1 and O_2 and agreement between the reports of the observers and the name of the exemplar.

Results

Table 2 shows how the major objectives of the tests were accomplished. The agreement between O_1 and O_2 was high for all seven tests; except for Moja, the agreement ranged between 86% and 95%. Note that this is the agreement for both correct and incorrect signs. The agreement between the signs reported by O_1 and O_2 and the correct names of the categories is also high; except for Moja, correct scores ranged between 71% and 88%.

The line labelled 'expected' in table 2 needs some explanation. When we first described this testing procedure (Gardner & Gardner 1971) we estimated the expected chance performance as $1/N$ where N is the number of vocabulary items on a test and all items are represented by the same number of exemplars. This estimate was based on the assumption that only the chimpanzees were guessing and that their guessing strategies could only be randomly related to the random sequence of presentation. But, this estimate may be too low because it does not

TABLE 1. VOCABULARY ITEMS IN THE TESTS OF FOUR CHIMPANZEES

items	chimpanzees W	M	T	D
animates				
BABY	+	.	+	+
BIRD	+	+	+	+
BUG	+	+	+	+
CAT	+	+	+	+
COW	+	+	+	+
DOG	+	+	+	+
HORSE	.	+	.	.
plants				
FLOWER	+	+	+	+
LEAF	+	+	.	.
TREE	+	+	+	+
clothing				
CLOTHES	+	.	.	.
HAT	+	+	+	+
PANTS	+	.	.	.
SHOE	+	+	+	+
grooming				
BRUSH	+	+	+	.
COMB	.	+	+	.
HANKIE	.	.	+	.
LIPSTICK	.	.	+	.
OIL	+	.	+	+
TOOTHBRUSH	+	+	+	+
WIPER	+	.	.	.
sensory				
LISTENS	+	+	.	.
LOOKS	+	+	+	.
SMELLS	+	.	.	.
foods				
APPLE	.	+	+	+
BANANA	.	+	+	+
BERRY	.	+	+	+
CARROT	.	.	+	+
CEREAL	.	+	.	.
CHEESE	+	.	+	+
CORN	.	+	+	+
FRUIT	+	.	.	.
GRAPES	.	+	.	.
GUM	.	+	.	.
ICE CREAM	.	+	+	+
MEAT	+	.	+	+
NUT	+	+	+	+
ONION	.	+	.	.
ORANGE	.	+	.	.
PEA-BEAN	.	+	.	.
PEACH	.	+	+	+
SANDWICH	.	.	+	+
TOMATO	+	.	.	.
drinks				
COFFEE	.	.	+	+
DRINK	+	.	.	.
MILK	.	+	.	.
SODAPOP	.	+	+	+
other				
BALL	.	.	+	+
BOOK	+	+	.	.
CAR	+	.	+	+
HAMMER	+	.	.	.
KEY	+	+	+	.
KNIFE	.	+	.	.
PEEKABOO	.	.	+	.
PIPE	+	.	.	.
SMOKE	+	+	.	.

take into account the possibility that the observers were guessing. In random sampling without replacement, the probabilities of later events in a sequence depend on earlier events. The observers could have used their knowledge of the items that had appeared earlier to predict the items that would appear later. Thus, players who can remember the cards that have already been played can win significant amounts at games such as blackjack. Diaconis (1978) and Read (1962) deal with a similar problem in demonstrations of extra-sensory perception (e.s.p.). When highly motivated subjects in e.s.p. experiments can see each target card after each prediction, their later predictions tend to improve.

To estimate the effect of informed guessing by the observers on chance expectancy in the tests reported here, James C. Patterson (1983) performed a computer simulation which assumes that both observers; (i) saw each slide after each trial; (ii) had perfect memory for the number of exemplars of each vocabulary item that had appeared before the beginning of each trial; and (iii) guessed the correct sign on the basis of the number of exemplars of each vocabulary

TABLE 2. SCORES ON THE VOCABULARY TESTS OF FOUR CHIMPANZEES

chimpanzee subject	Washoe		Moja	Tatu		Dar	
test	1	2	1	1	2	1	2
vocabulary items	16	32	35	25	34	21	27
trials	64	128	140	100	136	84	108
inter-observer agreement (%)	95	86	70‡	89	91	90	94
scored correct by							
observer 1 (%)	86	72	54§	84	80	79	83
observer 2 (%)	88	71	54§	85	79	80	81
expected† (%)	15	4	4	6	4	6	5

† Assuming that the observer was guessing on the basis of perfect memory for all previous trials that that observer had seen.

‡ Based on 135 trials; O_2 missed five trials.

§ Based on 132 trials; eight unscorable trials.

item that remained to be presented. The expected scores in table 2 are the average results of 2000 simulated runs for each of the seven tests. In all cases, this estimate is a small fraction of the obtained scores. Since, O_1 and O_2 reported extra-list intrusions (signs that were not on the target lists) they were using a less efficient strategy than that assumed in Patterson's simulation. Hence, small as they are, the values in the expected line of table 2 over-estimate chance expectancy.

The expected score for Washoe's first test is appreciably higher than the expected scores for the other six tests for two reasons. First, this test was shorter than the other tests and predictability depends on the number of vocabulary items: the fewer the items the greater the predictability. Second, and more significantly for this discussion, predictability increases as we approach the end of the test. The last trial is completely predictable, since there is only one vocabulary item that could have any remaining exemplars. The next to the last trial may be completely predictable, but there are at most two vocabulary items that could still appear, and so on. In all cases, except for Washoe's first test, both O_1 and O_2 were assigned to test sessions in such a way that no individual served as an observer for more than half of the trials of any single test. The device is similar to the way gambling casinos can defeat card-counting customers by reshuffling the deck. The smaller number of items and the assignment of the same two observers to both sessions of Washoe's first test account for the higher, but still quite small, expected score on that test.

Concepts

To make sure that the signs referred to conceptual categories, all of the test trials were first trials; that is, each slide was shown to the subject for the first time on the one and only test trial in which it was presented to that subject. All of the specific stimulus values varied, as they do in natural language categories: that is to say, most human beings would agree that the exemplars in each set belong together. Apparently, Washoe, Moja, Tatu, and Dar agreed with this assignment of exemplars to conceptual categories (See Gardner & Gardner 1984, figures 3, 4, 5 and 6).

When teaching a new sign, we usually began with a particular exemplar: a particular toy for BALL, a particular shoe for SHOE. At first, especially with very young subjects, there would be very few balls and very few shoes. The same situation is common in human nursery life. Early in Project Washoe we worried that the signs might become too closely associated with their initial referents. It turned out that this was no more a problem for Washoe or any of our other subjects that it is for children. The chimpanzees easily transferred the signs they had learned for a few balls, shoes, flowers, or cats to the full range of the categories wherever they found them and however represented, as if they divided the world into the same conceptual categories that human beings use.

It is reasonable to suppose that non-human animals use natural language concepts outside the laboratory. A wild monkey that finds a ripe mango in a tree must learn something general about ripe mangoes, because it is certain that that monkey will never get to pick that particular mango again. A young lion that brings down an impala must learn something general about hunting impalas, because it is certain that that lion will never get to hunt that particular impala again. The same must be true of young hawks hunting field mice. It seems unlikely that any creature with a natural world as complex as that of a wild pigeon could earn its living without using some natural language concepts.

So much experimentation has been limited to precisely repeated stimulus objects or to objects that vary only in simple dimensions, such as colour and size, that it would be easy to form the impression that the conceptual abilities of non-human beings are severely limited. But, there have been notable exceptions. Hayes & Hayes (1953) working with chimpanzee Viki, Hicks (1956) and Sands *et al.* (1982) working with monkeys, and Herrnstein *et al.* (1976) as well as Herrnstein (this symposium) with pigeons, for example, have all demonstrated that non-human beings can use natural language concepts when they are presented with suitably varied stimulus material.

Significant variation among exemplars and testing with true first trials are essential to the definition of natural language categories. More concerned with theoretical definitions of language than with reference, the Rumbaughs (Essock *et al.* 1977; Gill & Rumbaugh 1974; Savage-Rumbaugh *et al.* 1983) have administered hundreds of trials of training and testing with identical exemplars or with minimally varied exemplars. To be sure, the Rumbaughs have concentrated their efforts on the arbitrariness of what they call 'lexigrams' and the use of the 'lexigrams' in arbitrarily fixed sequences. The stated objective has been to satisfy certain theoretical definitions of 'symbolic communication', and such methods may be appropriate for that objective. It seems likely that, given the opportunity, the Rumbaugh chimpanzees could also have used natural language categories.

As for Terrace and his associates (Terrace 1979; Terrace *et al.* 1980) they never attempted any experimental analysis of reference. As a matter of fact their work is unique in this field in that they never administered any systematic tests at all. Since the time of Kellogg & Kellogg (1933), it is the first study in this field that was entirely restricted to adventitious naturalistic observations.

5. Communication and motive

Normal human children learn to speak as if they were born with a powerful motive to communicate; no extrinsic reward seems to be necessary. In modern times, of course, we recognize that there are many inborn motives rather than a few basic ones, such as hunger and thirst that give rise to the rest through a process of conditioning. Moreover, other inborn and unlearned motives, can be more powerful determinants of behaviour than hunger and thirst. Harry Harlow's experiments on contact comfort come immediately to mind (Harlow 1958). It is also clear now that many other species behave as if they were born with a powerful motive to communicate. The need to communicate is by no means uniquely human (Tinbergen

1953). Inborn motives such as contact, comfort and communication have obvious selective advantages. To the modern mind, the existence of many such inborn motives seems rather more compatible with Darwinism than the elaborate process of conditioning based on hunger and thirst that was formerly posited.

Chimpanzees are among the many species that behave as if they were born with a powerful motive to communicate (Van Lawick-Goodall 1968). Captive chimpanzees are similar to wild chimpanzees in this respect (Kellogg 1968) unless their conditions of captivity are so severe that normal behaviour is suppressed. On the basis of our own early observations (Gardner & Gardner 1971, p. 141), and reports of Viki, particularly Hayes & Nissen (1971) (written in 1957 and made available to us in draft form in the early days of Project Washoe) we learned to avoid all forms of drill. In the Reno laboratory, the only time that formal, trial-by-trial procedures were used for teaching A.S.L. (as opposed to testing) was in Fouts' (1972) experiment, a Ph.D. dissertation designed to isolate the effects of two specific procedures, modelling and moulding. Occasionally, we did introduce extrinsic rewards as, for example, when we rewarded Tatu and Dar with treats for obedient test-taking behaviour, but the extrinsic rewards usually had to be discontinued because their main effect was to interfere with the intrinsically motivated task at hand.

The following example taken from Hayes & Nissen (1971) is typical of cross-fostered chimpanzees.

> '...one hot summer day [Viki] brought a magazine illustration of a glass of iced tea to a human friend. Tapping it, she said 'Cup! Cup!' and ran to the refrigerator, pulling him along with her. It occurred to us that pictures might be used to signify needs more explicitly than words...
>
> A set of cards was prepared showing magazine illustrations in natural color of those things she solicited most frequently [for four days Viki consistently used the picture-cards for requests, but on the fifth day]...suddenly she acted as if imposed upon. She had to be coaxed to cooperate and then used the pictures in a completely random way.
>
> [after seven months of erratic performance]...the technique which had seemed so promising was dropped, pending revision. Spring weather, plus a new car, gave Viki a wanderlust so that no matter what situation sent her to the picture-communication pack, when she came upon a car picture she made happy noises and prepared to go for a ride. We eliminated all car pictures from the pack, but it was too late. Long afterwards Viki was tearing pictures of automobiles from magazines and offering them as tickets for rides' (pp. 107–108).

At the same time, experiments with human children (see, for example, Lepper *et al.* 1973; Levine & Fasnacht 1974) have demonstrated that the heavy-handed application of extrinsic rewards impairs performance on intrinsically motivated tasks, such as drawing. Heavy reliance on extrinsic rewards probably has a similar negative effect on the performance of chimpanzees. Characteristically, those who have relied most heavily on extrinsic rewards have been those who most insistently claimed that chimpanzees lack the intrinsic motivation to communicate (Savage-Rumbaugh & Rumbaugh 1978; Savage-Rumbaugh *et al.* 1983, pp. 462, 485–486; Terrace, 1979, pp. 221–224; Terrace *et al.* 1980, pp. 438–440).

Obviously, as their verbal skills improved, Washoe, Moja, Tatu, and Dar were more successful in making their wants known and, presumably, more successful in getting what they wanted. But, certainly the same can be said for human children, or human adults, for that matter. Also, those serving as O_1 often showed their approval or disapproval by smiling or frowning, by nodding or shaking their heads, and by praising the chimpanzees in A.S.L. This was carried over from the cross-fostering régime that was maintained at all times. It is the same way that human adults normally respond to the verbal behaviour of human children.

6. Concluding remarks

In this article we have reviewed a small portion of the published and still to be published record of our sign language studies of chimpanzees. In comparing these cross-fostered chimpanzees with human children, we must consider how very young and immature they were. Chimpanzees begin to lose their milk teeth when they are five or six years old. Under natural conditions in Africa, infants are not weaned until they are four or five years old, they usually live with their mothers until they are seven, and often continue to live with their mothers until they are ten or eleven. The youngest wild chimpanzee mother at the Gombe Stream was twelve years old when her first baby was born (Van Lawick-Goodall 1973). While the life span of wild chimpanzees is still unknown, we do know that captive chimpanzees can remain vigorously and intelligently alive for more than 50 years (Maple & Cone 1981).

Washoe was captured wild in Africa, so we cannot know just when she was born, but dentition and other indicators agree with our estimate of an age of nine to eleven months at the time she arrived in Reno. Washoe was maintained under cross-fostering for only 51 months, Moja only 78 months, Tatu only 65 months, and Dar only 58 months. At that age Moja, the oldest, had only lost six of her milk teeth. It is clear that they were much too young to demonstrate the limits of chimpanzee intelligence and the full benefits of cross-fostering. Moreover, these were the first subjects to be treated with the new technique. Our strongest conclusion is that there is still much more to be discovered about the continuity between verbal and non-verbal and human and non-human.

In the concluding remarks of our first publication on this subject, which covered Washoe's first 22 months of cross-fostering, we wrote,

> 'at an earlier time we would have been more cautious about suggesting that a chimpanzee might be able to produce extended utterances to communicate information. We believe now that it is the writers – who would predict just what it is that no chimpanzee will ever do – who must proceed with caution' (Gardner & Gardner (1969) pp. 671–672).

Bronowski & Bellugi (1970) were the first to publish an article on the limits of chimpanzee intelligence based on that partial report of Washoe's first 22 months in Reno. Each successive report of further progress has generated a fresh wave of commentators, and among the commentators there have always been those who were convinced that at last the final limits were in sight. In those circles, the new evidence has provoked, not caution, but ever more daring hypotheses about the theoretical basis for the newly discovered limits (Terrace *et al.* 1979; Terrace, this symposium).

In 1910, when aviation was probably more advanced than our cognitive sciences are today, two senior engineers, Jackman and Russell wrote in their pioneering book on the construction and operation of flying machines,

> 'in the opinion of competent experts it is idle to look for a commercial future for the flying machine. There is, and always will be, a limit to its carrying capacity which will prohibit its employment for passenger or freight purposes in a wholesale or general way. There are some, of course, who will argue that because a machine will carry two people, another may be constructed that will carry a dozen, but those who make this contention do not understand the theory of weight sustention in the air; or that the greater the load the greater must be the lifting power (motors and plane surface), and that there is a limit to these – as will be explained later on – beyond which the aviator cannot go' (p. 22).

Research supported by grants GB 7432, GB 35586, BNS 75-17290, BNS 78-13058, and BNS 79-13832, from the National Science Foundation; MH 12154 and MH 34953 (Research Development award to B. T. Gardner) from the National Institute of Mental Health; and by grants from the National Geographic Society, the Grant Foundation, and the Spencer Foundation. Dedicated to Niko Tinbergen, F.R.S., in appreciation of his teaching and of his leadership in blending observation and critical experimentation in ethology.

References

Bronowski, J. & Bellugi, U. 1970 Language, name and concept. *Science, Wash.* **168**, 669–673.

Brown, R. 1970 The first sentences of child and chimpanzee. In *Selected psycholinguistic papers* (ed. R. Brown), pp. 208–281. New York: Macmillan.

Diaconis, P. 1978 Statistical problems in ESP research. *Science, Wash.* **201**, 131–136.

Dingwall, W. O. 1979 The evolution of human communication systems. *Stud. Neuroling.* **4**, 1–95.

Donahoe, J. W. & Wessells, M. G. 1980 *Learning, language, and memory.* New York: Harper & Row Publishers.

Essock, S. M., Gill, T. V. & Rumbaugh, D. M. 1977 Language relevant object- and color-naming tasks. In *Language learning by a chimpanzee* (ed. D. M. Rumbaugh), pp. 193–206. New York: Academic Press.

Fant, L. 1977 *Sign language.* Northridge, California: Joyce Media, Inc.

Fouts, R. S. 1972 Use of guidance in teaching sign language to a chimpanzee. *J. comp. Physiol. Psychol.* **80**, 515–522.

Fouts, R. S., Fouts, D. H. & Schoenfeld, D. 1984 Sign language conversational interaction between chimpanzees. *Sign Lang. Stud.* **42**, 1–12.

Gardner, B. T. & Gardner, R. A. 1971 Two-way communication with an infant chimpanzee. In *Behaviour of nonhuman primates* (ed. A. Schrier and F. Stollnitz), vol. 4, pp. 117–184. New York: Academic Press.

Gardner, B. T. & Gardner, R. A. 1974*a* Comparing the early utterances of child and chimpanzee. In *Minnesota symposium on child psychology* (ed. A. Pick), vol. 8, pp. 3–23. Minneapolis: University of Minnesota Press.

Gardner, B. T. & Gardner, R. A. 1975 Evidence for sentence constituents in the early utterances of child and chimpanzee. *J. exp. Psychol.: General* **104**, 244–267.

Gardner, B. T. & Gardner, R. A. 1980 Two comparative psychologists look at language acquisition. In *Children's language* (ed. K. E. Nelson), vol. 2, pp. 331–369. New York: Gardner Press.

Gardner, R. A. & Gardner B. T. 1969 Teaching sign language to a chimpanzee. *Science, Wash.* **165**, 664–672.

Gardner, R. A. & Gardner, B. T. 1974*b* Teaching sign language to the chimpanzee, Washoe. *Bull. Audio Phonol.*, **4**, 145–173.

Gardner, R. A. & Gardner, B. T. 1978 Comparative psychology and language acquisition. *Ann. N.Y. Acad. Sci.* **309**, 37–76.

Gardner, R. A. & Gardner, B. T. 1984 A vocabulary test for chimpanzees. *J. comp. Psychol.* **98**, 381–404.

Gill, T. V. & Rumbaugh, D. M. 1974 Mastery of naming skills by a chimpanzee. *J. Human Evol.* **3**, 483–492.

Goodall, J. 1965 Chimpanzees of the Gombe Stream Reserve. In *Primate behavior: field studies of monkeys and apes* (ed. I. DeVore), pp. 425–473. New York: Holt, Rinehart & Winston.

Griffin, D. 1976 *The question of animal awareness: evolutionary continuity of mental experience.* New York: The Rockefeller University Press.

Harlow, H. F. 1958 The nature of love. *Am. Psychol.* **13**, 673–685.

Harris, M. P. 1970 Abnormal migration and hybridization of *Larus argentatus* and *L. fuscus* after interspecies fostering experiments. *Ibis* **112**, 488–498.
Hayes, K. J. & Hayes, C. 1953 Picture perception in a home-raised chimpanzee. *J. comp. Physiol. Psychol.* **46**, 470–474.
Hayes, K. J. & Nissen, C. H. 1971 Higher mental functions of a home-raised chimpanzee. In *Behavior of nonhuman primates* (ed. A. M. Schrier & F. Stollnitz), vol. 4, pp. 59–115. New York: Academic Press.
Herrnstein, R. J., Loveland, D. H, & Cable C. 1976 Natural concepts in pigeons. *J. exp. Psychol. Anim. Behav. Proc.* **2**, 285–302.
Hewes, G. W. 1973 Pongid capacity for language acquisition: An evaluation of recent studies. *Symp. Fourth int. Congr. Primatol.* **1**, 124–143.
Hicks, L. H. 1956 An analysis of number-concept formation in the rhesus monkey. *J. comp. Physiol. Psychol.* **49**, 212–218.
Hill, J. 1978 Apes and language, *Ann. Rev. Anthropol.* **7**, 89–112.
Hockett, C. F. 1978 In search of Jove's brow. *Am. Speech* **53**, 243–314.
Jackman, W. J. & Russell, T. H. 1910 *Flying machines: construction and operation.* Chicago: The Charles C. Thompson Co.
Kellogg, W. N. 1968 Communication and language in the home-raised chimpanzee. *Science, Wash.* **162**, 423–427.
Kellogg, W. N. 1969 Research on the home-raised chimpanzee. In *The chimpanzee* (ed. G. H. Bourne), vol. 1, pp. 369–392. Basel: S. Karger.
Kellogg, W. N. & Kellogg, L. A. 1933 *The ape and the child.* New York: Hafner.
Lenneberg, E. 1967 *Biological foundations of language.* New York: John Wiley.
Lepper, M. R., Greene, D. & Nisbett, R. E. 1973 Undermining children's intrinsic interest with extrinsic rewards. *J. Personal. soc. Psychol.* **28**, 129–137.
Levine, F. & Fasnacht, G. 1974 Token rewards may lead to token learning. *Am. Psychol.* **29**, 816–820.
Lieberman, P. 1984 *The biology and evolution of language.* Cambridge, Mass.: Harvard University Press.
Maple, T. L. & Cone, S. G. 1981 Aged apes at the Yerkes Regional Primate Research Center. *Lab. Primate Newsl.* **20**, 10–12.
Marler, P. 1969 Animals and man: communication and its development. In *Communication* (ed. J. D. Roslansky), pp. 25–61. Amsterdam: North-Holland.
McNeill, D. 1966 The creation of language by children. In *Psycholinguistics papers* (ed. J. Lyons and R. Wales), pp. 99–114. Edinburgh: Edinburgh University Press.
Moore, T. E. (ed.) 1973 *Cognitive development and the acquisition of language.* New York: Academic Press.
Patterson, J. C. 1983 Computational formulas for use in card-guessing with information. Paper presented at the Mid-South Colloquium in Mathematical Sciences, Memphis, Tennessee, February 1983.
Pfungst, O. 1911 *Clever Hans* (transl. by C. L. Rahn). New York: Henry Holt.
Premack, D. 1971 Language in chimpanzee? *Science, Wash.* **172**, 808–822.
Read, R. C. 1962 Card-guessing with information – a problem in probability. *Am. math. Monthly* **69**, 506–511.
Rosch, E. & Lloyd, B. B, (eds) 1978 *Cognition and categorization.* Hillsdale, N.J.: Lawrence Erlbaum.
Rumbaugh, D. M., Gill, T. V. & Von Glasersfeld, E. 1973 Reading and sentence completion by a chimpanzee. *Science, Wash.* **182**, 731–733.
Saltz, E., Soller, E. & Sigel, I. 1972 The development of natural language concepts. *Child Devel.* **43**, 1191–1202.
Saltz, E., Dixon, D., Klein, S. & Becker, G. 1977 Studies of natural language concepts. III. Concept overdiscrimination in comprehension between two and four years of age. *Child Devel.* **48**, 1682–1685.
Sands, S. F., Lincoln, C. E. & Wright, A. A. 1982 Pictorial similarity judgements and the organization of visual memory in the rhesus monkey. *J. exp. Psychol.: General* **111**, 369–389.
Savage-Rumbaugh, E. S. & Rumbaugh, D. M. 1978 Symbolization, language, and chimpanzees: A theoretical reevaluation based on initial language acquisition processes in four young *Pan troglodytes. Brain Lang.* **6**, 265–300.
Savage-Rumbaugh, E. S., Pate, J. L., Lawson, J., Smith, S. T. & Rosenbaum, S. 1983 Can a chimpanzee make a statement? *J. exp. Psychol.: General* **112**, 457–492.
Schlesinger, H. S. & Meadow, K. P. 1972 *Deafness and mental health: a developmental approach.* Berkeley: University of California Press.
Snow, C. 1977 Mother's speech research: from input to interaction. In *Talking to children* (ed. C. Snow and C. Ferguson), pp. 31–49. Cambridge: Cambridge University Press.
Stokoe, W. C. 1978 Sign language versus spoken language. *Sign Lang. Stud.* **18**, 69–90.
Stokoe, W. C., Casterline, D. C. & Croneberg, C. G. 1976 *A dictionary of American Sign Language on linguistic principles.* Silver Spring: Linstok Press.
Terrace, H. S. 1979 *Nim.* New York: Knopf.
Terrace, H. S., Pettito, L., Sanders, R. J. & Bever, T. G. 1979 Can an ape create a sentence? *Science, Wash.* **206**, 891–902.
Terrace, H. S., Pettito, L., Sanders, R. J. & Bever, T. G. 1980 On the grammatical capacity of apes. In *Children's language* (ed. K. E. Nelson), vol. 2, pp. 371–495. New York: Gardner Press.
Thorpe, W. H. 1972 The comparison of vocal communication in animal and man. In *Non-verbal communication* (ed. R. A. Hinde), pp. 27–47. Cambridge: Cambridge University Press.

Tinbergen, N. 1953 *Social behaviour in animals*. New York: John Wiley.
Van Cantfort, T. E. & Rimpau, J. B. 1982 Sign language studies with children and chimpanzees. *Sign Lang. Stud.* **34**, 15–72.
Van Lawick-Goodall, J. 1968 A preliminary report on expressive movements and communication in the Gombe Stream chimpanzees. In *Primates: studies in adaptation and variability* (ed. P. C. Jay), pp. 313–374. New York: Holt, Rinehart and Winston.
Van Lawick-Goodall, J. 1973 The behavior of chimpanzees in their natural habitat. *Am. J. Psychiat.* **130**, 1–11.
Watson, D. O. 1973 *Talk with your hands*. Menasha, Wisconsin: George Banta, Company, Inc.
Watt, W. C. 1974 Review of A. M. Schrier & F. Stollnitz (eds), *Behavior of nonhuman primates*. *Behav. Sci.* **19**, 70–75.
Yerkes, R. M. & Yerkes, A. W. 1929 *The great apes: A study of anthropoid life*. New Haven: Yale University Press.

Phil. Trans. R. Soc. Lond. B **308**, 177–185 (1985) [177]
Printed in Great Britain

The capacity of animals to acquire language: do species differences have anything to say to us?

By E. Sue Savage-Rumbaugh, Rose A. Sevcik, D. M. Rumbaugh and Elizabeth Rubert

Language Research Center, Yerkes Regional Primate Research Center, Emory University, Atlanta, Georgia, 30322, U.S.A.

Following the Gardners' discovery that an ape named Washoe could learn to produce and combine a number of hand movements similar to those used by deaf human beings, a variety of 'ape-language projects' sprang up. Some projects used different symbol systems, others used different training techniques, and others used different species of apes. While debate still rages regarding the appropriate way to interpret the symbolic productions of apes, three species of great apes (gorilla, orangutan, and chimpanzee) have now been credited with this capacity while no lesser apes or monkeys have been reported, at present, to have acquired such communicative skills. Among all of the claims made for the various animal species, the philosophers have entered the fray attempting to define the essence of what it is about language that makes it 'human'. This paper will compare and contrast the above positions to arrive at behavioural definitions of symbolic usage that can be applied across species. It will then present new data on a fourth ape species *Pan paniscus* which is proving to be the first non-human species to acquire symbolic skills in a spontaneous manner.

Ape-language began with what seemed to be a very simple and intriguing question: can apes learn to talk? This question has fascinated a number of psychologists ever since the discovery of apes by western civilization. The behaviour of these animals seemed so intelligent that many scientists were repeatedly puzzled as to why they could not learn to speak. A number of people tried to induce speech in young apes with relatively little success (see Kellogg 1968 for a review). The Gardners' breakthrough using a non-speech mode with Washoe was a major success (Gardner & Gardner 1975). Shortly thereafter, both Premack (1976) and Rumbaugh (1977), also using non-speech modes, reported linguistic breakthroughs. Once it became apparent, however, that chimpanzees were gesturing, using magnetic forms, and selecting lighted symbols with surprising agility, three major questions arose:

(i) what exciting thing will they say next?
(ii) How do we know if they know what they are saying?
(iii) What is language?

The first question was fuelled by popular accounts of the work, which appeared in newspapers and magazines regularly; each additional report sought to make some new and spectacular assertion. The second question, raised more slowly, initially came from academics who were somewhat incredulous regarding the initial claims. The third question was completely new. The human species had not previously bothered to determine exactly what language was and was not. Given that only human beings had language, it had not seemed especially important to determine which sorts of behaviours were to be judged 'linguistic' and which were not. With Washoe's first words, the definition of language suddenly became an issue of the first order.

Ape-language reports, regardless of the extent of language skills claimed for the apes, present both a challenge and a problem for psychology as a field. The problem is that none of the ape-language studies fits neatly into the extant categories of research. They are not 'animal learning' studies, in that they do not look for basic principles or for laws of learning; nor do they fit appropriately within the traditional matrix of ethologically based studies of animal communication, wherein a limited repertoire, consisting of a maximum of 100 'social signals' is the domain of study (Eibl-Eibesfeldt 1975).

While such technical issues as the presence or the absence of syntax and the role of conditioning are often the focus of public discussions of the phenomena, the critical issue that seems to fire the public imagination, and consequently fuel the research, is actually whether or not a species, other than man, can purposefully and consciously communicate, either among themselves or with humankind. Can they really tell us how they feel and what they think? Can we ask them questions? Can they ask us questions? And if they can talk, what sorts of things will they say? Most people refuse to believe that the answers to these questions should be complicated; after all, would not everyone know it if their dog suddenly began talking to them? Why then should scientists have a difficult time deciding what it is apes are doing?

These seemingly simple questions lead deep into the heart of psychology, to the point where it merges with philosophy. As psychologists, we have not clearly determined what is involved in 'telling someone something', nor do we know exactly how to ascertain that it is actually happening since 'telling' implies a special set of events between two entities. Is our dog 'telling' us that he wants to go out when he scratches at the door, or is his scratching 'conveying' to us that he wants to go out? It makes a difference. 'Telling' involves behaviour that is mediated not by the self, but by the actions of others; here one brings about consequences indirectly, by talking, or by conveying a message, as opposed to executing a direct action.

Thus, it must be recognized that the real issues involved in ape-language go far deeper than words and syntax. They involve the very nature of inter-individual relationships as we as human beings know and experience them, for a most distinctive human characteristic is that we do 'tell' each other things, whether it be in words, gestures, or pictures drawn in the dirt. We do knowingly, and with intent 'tell' each other things that allow us to transmit indirect experiences from generation to generation, to produce myth from fact, to build and to maintain unique cultures, and to know other human minds in ways that we do not know the minds of other species on this planet. It is no small thing, therefore, to assert that apes are 'telling' people things, or that they are 'telling' each other things.

Confusion between human and animal communication often arises over the application of different paradigms to the study of human behaviour, as opposed to animal behaviour. The paradigm problem is compounded and amplified by the restriction of specific terms to the description of human communicative patterns alone. For example, when used in reference to human communication, the term 'awareness', implies something that is not generally attributed to animals. That 'something' is the human knowledge that communicative acts are not simply another form of behaviour, rather they are *behaviours about intended future alterations in the behaviours of others*, or 'verbal behaviours'.

An easy way to visualize the distinction between behaviours that exist in their own right and behaviours that are 'verbal', or about the behaviour of others, is to consider the prespeech 'conversational babbling' of human infants. During such 'conversational exchanges' between mother and infant, there is a reciprocal exchange of roles and the intonation patterns of the infant typically follow those of the mother in a responsive manner. At this stage, however, it appears that the infant is not, in fact, saying anything, rather he or she is producing the vocal exchange as a behaviour in its own right. Later on, when the infant is several months older, the mother will comment that the infant now knows what he or she is saying (see, for example, Locke 1978, 1980). At this stage, the vocalizations may still be unintelligible to a novice observer, and they may even be unintelligible to the mother herself, yet as a result of the surrounding behavioural context, the vocalizations will be recognized as intentional communications, albeit rather poor in quality. That is, the act of uttering alone will no longer fulfil its own function; the infant will insist that specific sorts of action be taken in response and thus utterances will be judged to have become behaviours about behaviours, or intentional communicative acts.

Along with the appearance of such intentional communicative acts comes the concomitant awareness that other individuals also produce verbal behaviours and that they too make choices about engaging in a behavioural response to the verbalization of others. Thus, others may offer a response to the infant's communicative acts or elect not to do so. If they do not respond, the infant may perceive their lack of response as resulting from one of two events: either the other party did not understand nor perceive his or her intentional communication, or the other party received the communication, but chose not to respond. When the adult chooses not to respond, it will typically be made apparent to the youngster because the adult will first acknowledge receipt of the communication and then indicate his or her unwillingness to comply. Premack & Woodruff (1978) have termed the ontogenetic appearance of intentional communicative capacity, the achievement of a 'theory of the mind', meaning the development of the capacity to attribute mind to other individuals.

Nothing quite so esoteric as the mind, however, needs to be evoked. What we do need to suppose is the evolutionary advance of an ever-increasing ability to monitor the results of one's actions. First, only immediate effects would be monitored, then more indirect long term effects, and eventually the effects of one's own behaviour on the behaviours of others. Finally, we would begin to have an ability not only to monitor the effects of one's own actions, but also the effects of the actions of others. From this capacity would grow quite naturally a desire to control the actions of others for personal ends, and from this desire a need for a communicative system capable of representing to others the actions one would have them take.

Generally, one can determine the effectiveness of one's own behaviour by the ensuing events which affect the individual directly. Intentional communicative acts, however, must be monitored by judging a change in the behaviour of another and determining whether or not that change corresponds with the alteration which the intentional communicative act was intended to achieve. The phylogenetic onset of this skill remains to be adequately traced as strong evidence for this capacity is presently extant only in apes. It is possible, however, that species with highly developed brains in other orders, such as the Delphinidae, may also have developed this level of intentional communicative capacity. Observations of these animals in an interspecies communicative setting strongly suggests this (L. Herman, personal communication).

We wish to assert here that:

(i) apes are capable of intentionally telling one another things, and probably do so more in the wild than we yet realize;

(ii) apes who have been taught to communicate with human beings, and to use an arbitrary symbol system can, and do, tell each other, and their human companions, far more than apes who have not been so trained;

(iii) the type of training which the apes receive determines the degree to which their communicative behaviour becomes symbolic and abstract;

(iv) very large species differences with regard to communicative capacities exist between even the two most closely related great apes, *Pan troglodytes* and *Pan paniscus*.

At the Language Research Center, a joint venture of Georgia State University and the Yerkes Regional Primate Research Center of Emory University, we are trying to develop and to teach a viable system that permits individuals to interact communicatively and cooperatively with one another. In our efforts to teach language to *Pan troglodytes*, we have found that we have learned as much, if not more, from our failures as from our successes. The chimpanzees' halting acquisition of symbols has taught us that language is not made of whole cloth, but of many pieces. It is not merely symbols, or combinations of symbols, but complex ways of interacting, and complex sorts of inter-individual expectancies that intertwine and coordinate behaviour. Symbols are the medium of language, but not its substance. Its substance is planned, coordinated cooperation achieved only through mutual telling and mutual expectancies regarding that telling. These expectancies must be shaped by common past experiences and such experiences must occur in an atmosphere of trust and cooperation. Language, in short, is a part of culture; it is living, breathing, ever-changing, behavioural culture. It does not leave tangible artefacts, but it changes the behaviour of groups in a drastic manner. It cannot belong to one person; it is the property of inter-individual interactions. Without such interactions, language, as we know it, does not exist.

Our goal of achieving cooperative–communicative behaviour between chimpanzees has been, in a sense, a piecemeal approach to language that has attempted to provide chimpanzees with symbol use skills not already in their repertoire, and to integrate these skills with general behaviours.

We have, bit by bit, dissected the symbol and found it to be composed of a variety of skills which can be and must be taught individually if common chimpanzees are to come to be able to produce anything like human words. Once these individual skills are acquired, they can be integrated in a manner which permits the emergence of true indicative behaviours. We did not intend to begin with such a piecemeal approach as we set out to determine whether or not apes can acquire human-like language skills, yet, we repeatedly found that teaching 'names' was not sufficient and that the communications that resulted seemed to lack certain fundamental properties generally found in human communication. (For a complete analysis of the subskills that proved necessary to develop independently (requesting, naming, comprehending and stating) see Savage-Rumbaugh (1984)).

True indicative word-use involves the skill of 'telling' another, by means of a symbol, what you are going to do, what you see, what you know, etc., as opposed to simply emitting a gesture or selecting a symbol to obtain that which you want and could easily take, if but permitted. We now know that our initial view of what counted as a 'word' in a chimpanzee's vocabulary was naive. It has become clear that knowing that a name stands for something is far more complicated than simply being able to make the correct response in its presence. Far more crucial to language is the ability to predict or to inform, that is, to talk about what you are going to do before you do it, or what you want when you cannot see it. It is also procedurally more difficult to show that this is in fact what a chimpanzee has learned to do.

We have recently completed a series of blind tests which evaluated Sherman and Austin's ability to produce an indicative statement, that is, to symbolically encode their own actions on objects before the action occurs. (Savage-Rumbaugh *et al.* 1983; see figure 1).

By then evaluating the concordance between what they said they were going to do, and what they actually did, we obtained a measure of their capacity to use symbols in an indicative manner, that is, to indicate to another what it was they were about to do (see Table 1).

Before this test, this behaviour had become a very common part of Sherman's and Austin's spontaneous symbol use capacity. They had begun to say for example, 'GO SINK' and head for the kitchen, or 'SHERMAN M&MS' and drag the bag of candies out of the refrigerator. They had also begun to inform us of things they saw, as when Austin observed an anaesthetized chimpanzee carried past the window and said 'SCARE' while making 'waa' calls toward the window and gesturing for us to look out.

More recently, we have had the opportunity to study a completely new species of ape, *Pan paniscus*. This species has not before been the subject of language studies though it has attracted significant scientific attention for many other reasons (see, for example, Savage *et al.* 1977). It has also been proposed that *Pan paniscus* is more representative of the more generalized anatomical characteristics displayed by human ancestors than other living apes (that is, longer legs, shorter arms, reduced prognathism, etc.) (Zihlman *et al.* 1978).

We have been working with a young male, Kanzi, who was initially exposed to human language at six months old. He remained, however, with his mother until two and a half years old; during the time he was in his mother's company, he showed little interest in using the geometric symbols to communicate, though he was highly communicative gesturally and was exceedingly vocal, even attempting to produce some vocal imitations of speech in terms of pitch and intonation.

When Kanzi was two and a half years old he was separated from his mother so that she might return to the *Pan paniscus* social group and again become pregnant. Kanzi remained at the Language Research Center.

Kanzi's use of the keyboard became prominent immediately following his mother's absence and his symbol usage was completely spontaneous, that is, no training was given. Within a week of his mother's absence, it was determined that Kanzi could proficiently use many of the symbols which we had been attempting to teach his mother during the previous year. Moreover, not only could he use these symbols to request things he desired, he also demonstrated an ability to name things he did not immediately want and he displayed symbolic comprehension when others attempted to use the keyboard to communicate with him. Each of these skills had required separate training in the case of Sherman, Austin, and Lana. Kanzi's linguistic abilities were so astounding and strikingly different from those demonstrated by Sherman and Austin, and he had acquired them with no direct instruction, that the planned programme of instruction for Kanzi was completely revised. We decided that no moulding or training of any sort would be done with Kanzi at any time. In addition, no activity or object would be made contingent upon symbol usage, as had been the case with all previous ape-language studies, both our own, and those of others in the field. Kanzi's symbol usage, from the start, has been quite different from that of other apes. He uses symbols much more spontaneously and frequently directs our attention to places and to things that are not visible or to activities in which we are not, at present, engaged (see figure 2).

For Kanzi, the lexigram keyboard is a means of telling us that he wants to visit the A-FRAME,

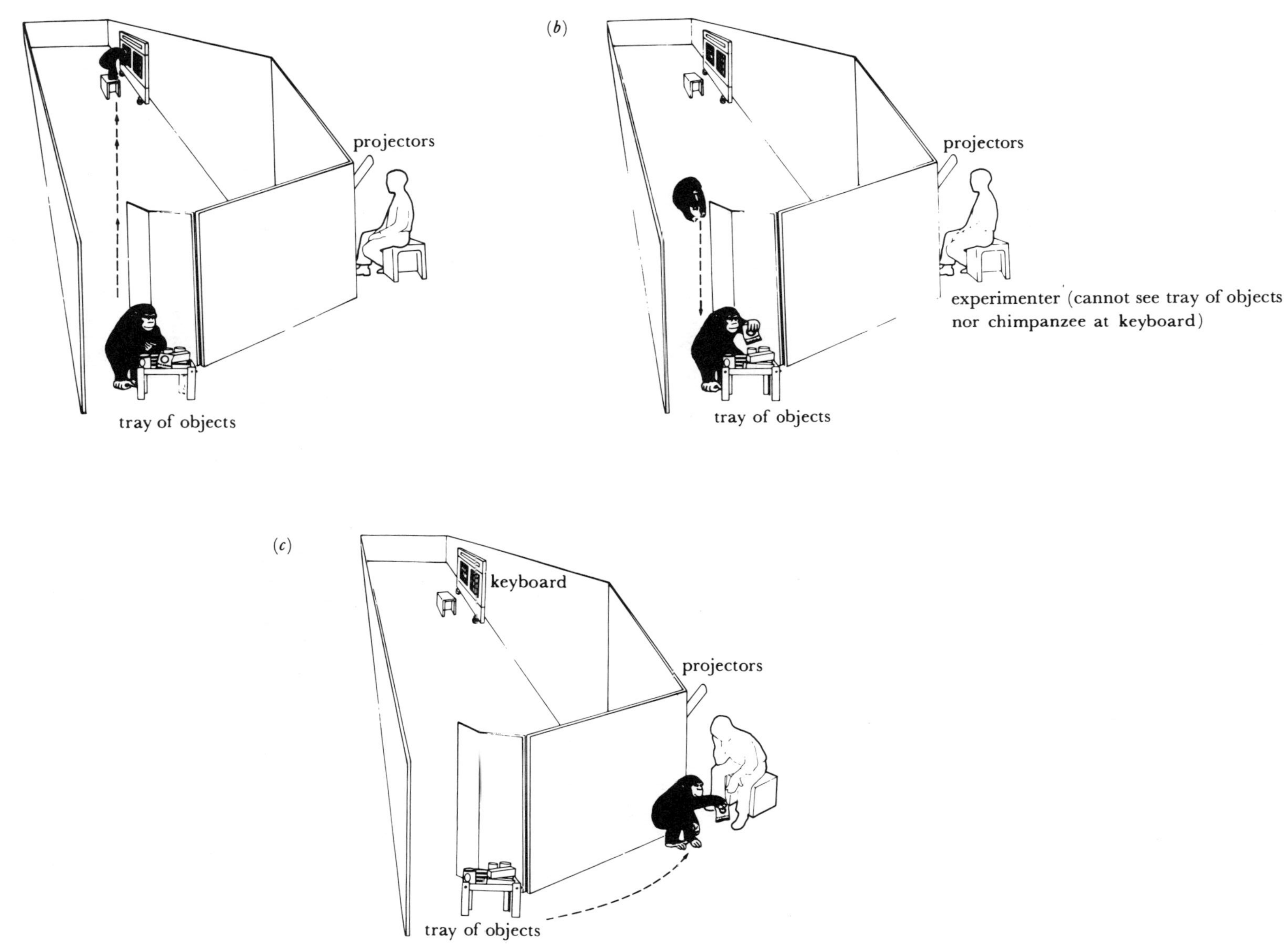

FIGURE 1. (*a*) The chimpanzee scans a tray of objects to see the available array. He returns to the keyboard and activates a symbol. While at the keyboard, the tray is out of the chimpanzee's view; he must therefore recall one or more of the items viewed on the tray.

(*b*) The chimpanzee then returns to the tray of objects and selects the item that he has activated on the keyboard. At this point, he must recall the symbol or object he has indicated on the keyboard since he can no longer view the display.

(*c*) The chimpanzee transports the object to the experimenter. The experimenter has been out of view during the chimpanzee's selection of the object. The experimenter refers to the projectors to verify whether the object delivered to him or her is, in fact, the one that the chimpanzee indicated at the keyboard.

TABLE 1. TOTAL SCORE AND PERCENTAGE CORRECT OF INDICATIVE STATEMENTS UNDER BLIND TEST CONDITIONS

	total score	percentage correct
Sherman	50/53	94 %
Austin	46/53	87 %

To obtain a correct score, the object that the chimpanzee selected and gave to the teacher had to correspond with the object the chimpanzee said that it would give the teacher. The array of objects changed in each trial.

FIGURE 2. Kanzi, a male infant *Pan paniscus*, communicates using a lexigram keyboard.

to go see the DOGS, to go where we keep the M&MS, or to find his beloved BALL. Symbol usage in Kanzi's case has not appeared piecemeal; he has not integrated training in many different situations and finally begun to make statements and to use his symbols representationally. Rather, he has displayed all of these skills from the outset quite spontaneously. By the end of the first six months of his mother's absence we had assigned symbols to 60 different keys on his board and he was using nearly half of them appropriately and spontaneously. He would also name or give photographs in response to a teacher's request, thereby enabling us to test his knowledge apart from a communicative context. No food reward nor training was necessary to interest Kanzi in taking such 'tests'. Moreover, he began spontaneously producing combinations of lexigrams and combining lexigrams with gestures, and in some cases with vocalizations. In contrast to Sherman and Austin, he also nearly always responds to our vocal questions with some sort of vocalization of his own, as though he is attempting to 'talk'. Although he cannot produce English words clearly, he does regularly use an 'aaangh' sound as a means of emphasizing 'I want' and uses other sounds to indicate denial or affirmation.

The most readily discernible and profound difference between Kanzi and Sherman and Austin seems to lie in Kanzi's ability to comprehend, at a still undetermined level, spoken English. Although others have claimed that chimpanzees comprehend spoken English, they have failed to present adequate data to substantiate these assertions (Fouts *et al.* 1976). In repeated tests since 1977, Sherman and Austin have consistently failed comprehension tests of spoken English though they constantly have been exposed to it from infancy. Kanzi, by contrast, is displaying marked comprehension of spoken English (see table 2).

TABLE 2. TOTAL SCORE OR PERCENTAGE CORRECT OF RECEPTIVE TRIALS

	stimulus condition			
	lexigram		spoken english	
	total score	percentage correct	total score	percentage correct
Kanzi	88/93	95%	83/93	89%
Sherman	92/93	99%	47/93	50%
Austin	91/93	98%	55/93	59%

Two (or three) exemplars were available for response on each trial. A total chance score would be 43%.

Early observations of this ability came as we asked him to name an item at his keyboard. (Recall that he was not being reinforced nor trained to do this, merely asked to do it if he could and if he would.) At times he seemed to have trouble finding a lexigram symbol, and we would say 'Kanzi, can't you find apple?'. Shortly after we uttered the word 'apple' in English, Kanzi's eyes would brighten and he would immediately touch the 'APPLE' lexigram on the keyboard. This behaviour was exceptionally striking, since no one who worked with Austin and Sherman could ever recall a single instance in which saying the word in English had helped them find it on the keyboard, even though we had said words many thousands of times across the past nine years of working with them. In no way was our use of English around Kanzi any different than it had been with them. English has been used freely with all of the chimpanzees in combination with the keyboard and with spontaneous gestures.

The symbols and sentences used by Kanzi are wholly learned; yet they are not conditioned through training any more than the words and the symbols used in this paper are so conditioned. The satisfactory explanation of Kanzi's linguistic competency, and its contrast with *Pan troglodytes*, present the current psychological and philosophical views of mankind with a unique and an unprecedented challenge.

This research was supported by N.I.H. grants NICHD-06016 and RR-00165 to The Yerkes Regional Primate Research Center, Emory University, Atlanta, Georgia, U.S.A. The authors wish to thank Jeannine Murphy, Kelly McDonald and Linda Bolser for their invaluable assistance in the data collection.

REFERENCES

Gardner, B. T. & Gardner, R. A. 1975 Evidence for sentence constituents in the early utterances of child and chimpanzee. *J. exp. Psychol.: General* **104**, 244–267.
Eibl-Eibesfeldt, I. 1975 *Ethology. The biology of behavior*, 2nd edn. New York: Holt, Rhinehart and Winston.
Fouts, R. S., Chowin, W. & Goodin, L. 1976 Transfer of signed responses in American Sign Language from vocal English stimuli to physical object stimuli by a chimpanzee (*Pan*). *Learning Motiv.* **7**, 458–475.
Kellogg, W. 1968 Communication and language in the home-raised chimpanzee. *Science, Wash.* **162**, 423–427.
Locke, A. (ed.) 1978 *Action, gesture and symbol: The emergence of language*. London: Academic Press.
Locke, A. 1980 *The guided reinvention of language*. London: Academic Press.
Premack, D. 1976 *Intelligence in ape and man*. Hillsdale, New Jersey: Lawrence Erlbaum Associates.

Premack, D. & Woodruff, G. 1978 Does the chimpanzee have a theory of mind? *Behav. Brain Sci.* **4**, 515–526.

Rumbaugh, D. M. (ed.) 1977 *Language learning by a chimpanzee: The LANA Project.* New York: Academic Press.

Savage, E. S., Wilkerson, B. J. & Bakeman, R. 1977 Spontaneous gestural communication among conspecifics in the pygmy chimpanzee (*Pan paniscus*). In *Progress in ape research* (ed. G. H. Bourne), pp. 97–116. New York: Academic Press.

Savage-Rumbaugh, E. S. 1984 Verbal behavior at a procedural level in the chimpanzee. *J. exp. Analys. Behav.* **41**, 223–250.

Savage-Rumbaugh, E. S., Pate, J. L., Lawson, J., Smith, S. T. & Rosenbaum, S. 1983 Can a chimpanzee make a statement? *J. exp. Psychol.: General* **112**, 457–492.

Zihlman, A., Cronin, J. E., Cramer, D. L. & Sarich, V. M. 1978 Pygmy chimpanzee as a possible prototype for the common ancestor of humans, chimpanzees and gorillas. *Nature, Lond.* **275**, 744–746.

Phil. Trans. R. Soc. Lond. B **308**, 187–201 (1985) [187]
Printed in Great Britain

Social and non-social knowledge in vervet monkeys

BY D. L. CHENEY AND R. M. SEYFARTH
Department of Anthropology, University of California, Los Angeles, California 90024, U.S.A.

The social knowledge of East African vervet monkeys is striking. Within a local population the monkeys recognize individuals, and associate each individual with its particular group. Within groups, the monkeys recognize dominance relations, rank orders, and matrilineal kinship, and they remember who has behaved affinitively towards them in the past. Outside the social domain, however, vervets appear to know surprisingly little about other aspects of their environment. Although they do distinguish the different alarm calls given by birds, vervets do not seem to recognize the fresh tracks of a python, or indirect evidence that a leopard is nearby. Similarly, although cooperation and reciprocity seem common in social interactions, comparable behaviour has apparently not evolved to deal with ecological problems. Results support the view that primate intelligence has evolved mainly to solve social problems. As a result, vervet monkeys make excellent primatologists but poor naturalists.

INTRODUCTION

We think of ourselves as relatively intelligent creatures for at least two reasons. First, humans not only learn rapidly when taught, they also acquire information without active instruction, by observing objects and events in the world around them. Secondly, human intelligence is not domain-specific, and knowledge acquired in one domain can readily be applied to another. Thus, a principle derived from a social problem can easily be applied to a logically similar problem involving objects.

Comparing human intelligence with the intelligence of non-human primates is difficult, because primate intelligence has thus far been measured almost exclusively by performance on learning tests. Comparatively little is known about the knowledge that monkeys and apes acquire naturally, in the absence of human intervention. More importantly, animal intelligence is generally tested only in one domain, using biologically arbitrary objects as stimuli. Most of the problems confronting non-human primates under natural conditions, however, are ones that derive from competitive and cooperative interactions with conspecifics. There is reason to believe that primates may reveal greater intelligence when dealing with each other rather than with irrelevant objects. In this paper, we examine what free-ranging vervet monkeys have learned, without human intervention, about their environment. We do so by means of observations and experiments that attempt to compare primate performance in social and non-social domains.

Primates tested in the laboratory, with objects, often face problems that are logically similar to the social problems confronting primates in the wild. Despite this similarity, however, the performance of primates in these two settings often seems to differ strikingly. To cite just one example, McGonigle & Chalmers (1977) and Gillan (1981) demonstrated transitive inference in captive squirrel monkeys and chimpanzees, respectively, but were able to do so only after considerable training with paired stimuli. In contrast, field observations suggest that, even from a very young age, monkeys are readily able to deduce a dominance hierarchy among conspecifics from their observation of dyadic interactions (Cheney 1978; Seyfarth 1981; Datta 1983; Gouzoules *et al.* 1984). Observations and experiments have also suggested that primates regularly classify individuals on the basis of kinship or close association (for example, Bachmann & Kummer 1980; Cheney & Seyfarth 1982*a*; Judge 1982; Smuts 1985). Moreover, numerous examples from field studies (admittedly anecdotal) suggest that primates can predict the consequences of their own actions on others, and understand enough about the behaviour and motives of others to be capable of deceit, and other subtle forms of manipulation (for example, Goodall *et al.* 1979; de Waal 1982; Kummer 1982; Cheney & Seyfarth 1984). Such observations are both intriguing and frustrating, because they suggest the existence in the wild of striking mental abilities that, with some notable exceptions (Woodruff & Premack 1979; Premack & Premack 1982), have not been documented or duplicated in the laboratory.

Because of the qualitative differences in field and laboratory stimuli, one possible explanation for the animals' differing performance suggests that primate intelligence is relatively domain-specific. This hypothesis argues that group life has exerted strong selective pressure on the ability of primates to form complex associations, make transitive inferences, and predict the behaviour of fellow group members. Thus abilities that seem to emerge only with human training in captivity may readily occur in primates under natural conditions, but mainly in the social domain (Jolly 1966; Chance & Jolly 1970; Humphrey 1976; Kummer 1971, 1982; see also Rozin 1976). Similarly, when captive chimpanzees solve technological problems that require foresight and an understanding of the consequences of past decisions (Dohl 1968), they may be demonstrating abilities for which they have been preadapted as a result of the need to make equally strategic decisions about each other (de Waal 1982).

This domain-specific hypothesis posits that natural selection may have acted to favour complex abilities in the social domain that are, for some reason, less easily extended or generalized to other spheres. The hypothesis does not, however, specify exactly how elaborate a monkey's social knowledge is, or what processes underlie it, nor does it claim that social knowledge can never be extended to other spheres. It argues simply that certain problems are solved more easily in the social domain as compared with other areas.

An alternative view argues that the contrast in ability between social and non-social behaviour derives not from any fundamental difference in ability between the two domains, but from the animals' lack of motivation to perform under laboratory conditions. Research on captive primates has been plagued by motivational problems, and it has often been difficult to distinguish between a lack of ability and a lack of incentive to perform the task at hand (for example, Terrace *et al.* 1979).

We have begun an investigation of these hypotheses by presenting free-ranging vervet monkeys with logically similar problems involving 'social' and 'non-social' stimuli. Although our field experiments are less precisely controlled than laboratory tests, they have at least two advantages. First, problems of motivation and human training are circumvented. Second, free-ranging primates daily encounter similar social and non-social problems, thus permitting a direct test of performance in the two domains. Third, our subjects regularly deal with objects in the external world that may be either relevant or irrelevant to their survival. It is therefore possible to compare social knowledge both with non-social knowledge of biologically relevant objects and with non-social knowledge of objects that are apparently unrelated to the animals' survival.

At the outset, two points should be emphasized. First, throughout the paper we draw a distinction between the performance of primates in the 'social' and 'non-social' domains. While we believe that this distinction is a real and heuristically important one, we recognize that the boundary between these spheres of activity is ill-defined. Secondly, in evaluating observations and experiments, we make no claims about the mechanisms underlying performance. Our experiments define knowledge operationally; they measure only the responses that particular stimuli evoke, and not the processes (mental or otherwise) that underlie such responses. Many of the results we describe could, for example, result either from relatively simple associative learning or from more complex cognitive processes. Our aim is not to argue for one of these alternatives. Instead, we use experiments to determine which of two stimuli is more salient, and to suggest that animals form some sorts of associations more readily than others.

1. Study site and subjects

Experiments were conducted on three free-ranging groups of vervet monkeys in Amboseli National Park, Kenya. Vervet monkeys live in stable social groups consisting of a number of adult males, adult females, and their juvenile and infant offspring (Cheney *et al.* 1985). As in most Old World monkeys, female vervets remain in their natal group throughout their lives, maintaining close bonds with maternal kin. Males, in contrast, emigrate to neighbouring groups at sexual maturity, often in the company of brothers or natal group peers (Cheney & Seyfarth 1983). Within each group, males and females can be ranked in linear dominance hierarchies that predict the outcome of competitive interactions over food, water and social partners. Offspring acquire dominance ranks immediately below those of their mothers, such that all members of a family typically share adjacent ranks (Cheney 1983*a*).

2. Relevant aspects of other species' behaviour

Vervets in Amboseli are preyed upon by leopards, a number of small carnivores, two species of eagle, baboons and pythons (Cheney & Seyfarth 1981). Predation is a major cause of mortality (Cheney *et al.* 1981, 1985). The monkeys give acoustically distinct alarm calls to different predators, and experiments have shown that each of these alarms evokes qualitatively different escape responses (Seyfarth *et al.* 1980). Calls given to leopards, for example, cause monkeys to run into trees, while calls given to eagles cause monkeys to look up in the air. The monkeys' alarm calls therefore function to designate different types of danger in the external world.

Vervets are not the only species to give alarm calls to predators, however, and it would seem advantageous for the monkeys to distinguish among alarm calls given by other species. We investigated the vervets' knowledge of other species' alarm calls through playback experiments with three different calls of the superb starling (*Spreo superbus*). Starlings give two acoustically distinct alarm calls to predators, neither of which bears any acoustic resemblance to the vervets' own alarms. One starling alarm – a harsh, noisy call – is given to various terrestrial predators (including vervets), all of which prey on starlings or their eggs but only some of which prey on vervets. The second alarm – a clear, rising tone – is given to many species of hawks and eagles, two of which prey on vervets.

In conducting playback experiments of starling alarm calls, we followed the same protocol previously used in tests of the vervet's own alarm calls (Seyfarth *et al.* 1980). First, we hid a loudspeaker near a group of one to five vervets ($\bar{x} = 17.4$ m; s.d., 5.3). The monkeys were then filmed for 10 s, to establish the probability that they would show a given response in the absence of any call. We then played one of the starlings' calls, and continued to film the monkeys' responses for another 10 s. Three starling calls were used: their ground predator alarm call, their aerial predator alarm call, and, as a control, their song. Individual monkeys generally appeared only once in all trials, and successive experiments on the members of a given social group were always separated by at least 48 h. Results are presented in figure 1.

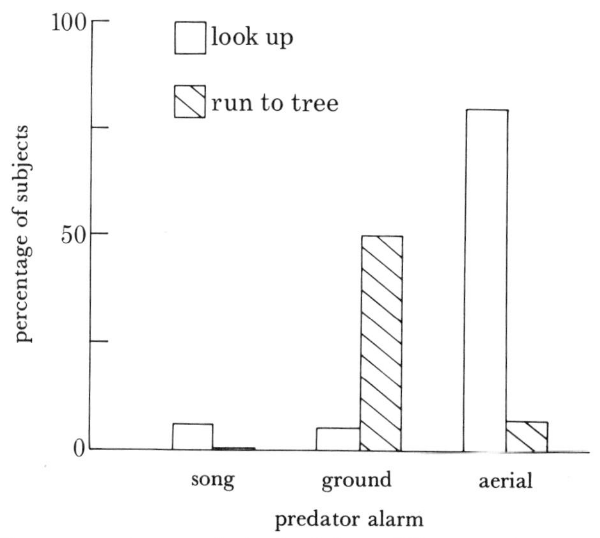

FIGURE 1. Responses of vervet monkeys to playback of three different starling vocalizations. Number of subjects for playback of song, terrestrial predator alarm, and avian predator alarm were 17, 18 and 15, respectively. Ground predator alarms evoked significantly more running to trees than either song ($\chi^2 = 11.4$, $p < 0.01$) or avian predator alarms ($\chi^2 = 7.3$, $p < 0.04$); avian predator alarms evoked significantly more looking up than either song ($\chi^2 = 18.2$, $p < 0.01$) or terrestrial predator alarms ($\chi^2 = 19.0$, $p < 0.01$).

Playback of the starling's ground predator alarm caused a significant number of monkeys to run into trees, while playback of the aerial predator alarm caused a significant number of vervets to look up. In contrast, the starling's song elicited little response. In this test, therefore, where the behaviour of another species was relevant to the vervets' survival, the monkeys' knowledge of another species' calls was similar to their knowledge of their own calls.

3. Apparently irrelevant aspects of other species' behaviour

The alarm calls of other species represent one end of a continuum of biologically relevant or irrelevant stimuli in the external world. It is perhaps not surprising that vervets discriminate between such alarm calls, since these calls are so obviously important to their survival. Can similar knowledge be demonstrated, however, for aspects of another species' behaviour that are apparently unrelated to the monkeys' survival? This seems an important question, because one striking feature of human intelligence is our inclination to accumulate information about the world that is not directly relevant to our survival. Can the same be said of vervet monkeys? Are vervets as good naturalists as they are primatologists?

To address this question one must first identify two comparable features of the monkeys' environment, one social and biologically relevant, the other non-social, and apparently irrelevant to the monkeys' survival. Then experiments must be designed to compare the monkeys' knowledge in these two domains. As a social, biologically relevant test, we asked the monkeys how much they knew about the ranging behaviour of other vervets. As a non-social, apparently irrelevant test we asked the monkeys how much they knew about the ranging behaviour of other species that neither compete nor interact with vervets in any obvious way.

Vervet monkeys aggressively defend their group's range against incursions by other groups. Females and juveniles are active participants in intergroup encounters, and give a distinctive vocalization when they spot the members of another group (Cheney 1981; Cheney & Seyfarth 1982*a*). In testing the vervets' knowledge of other groups' membership and ranges, subjects in one group were played the intergroup call of an animal from a neighbouring group, either from the true range of the vocalizer's group or from the range of another neighbouring group. In these paired trials, subjects responded with significantly more vigilance to calls played from the 'inappropriate' range than to calls played from the 'appropriate' range (Cheney & Seyfarth 1982*a*).

Subsequent experiments followed the same design, but used as stimuli the calls of other species. Vervets were played the calls of two species that are habitually found in or near water, the hippopotamus and the black-winged stilt (*Himantopus himantopus*). The hippopotamus's call is a territorial call, while the black-winged stilt's is a low-intensity alarm given to a wide variety of potentially disturbing species. These two species were chosen because neither competes nor interacts with vervets, and both are therefore of little biological importance to the monkeys. Nevertheless, each is a species that is so restricted to wet areas during the day that any indication of its presence in another habitat might be regarded, at least by humans, as anomalous. Black-winged stilts are never found away from water, and although hippos do emerge from water to feed on dry land, they do so only at night (Olivier & Laurie 1974).

Hippo and stilt calls were played to vervets either from the edge of a swamp or from a dry woodland area that contained no permanent water. All subjects were members of groups whose ranges bordered both types of area, and all had regularly heard the calls of both hippos and black-winged stilts when foraging near the swamp. Subjects were played hippo or stilt calls in paired trials, from either the swamp ('appropriate') or dry woodland ('inappropriate') habitat. Hippo calls were played at a mean distance of 91.9 m (s.d., 18.3) from the subjects, while stilt calls were played at a mean distance of 40.4 m (s.d., 11.2). As in previous experiments, these distances reflected the different calls' relative amplitudes. Order of presentation was systematically varied, and no individual appeared as a subject in more than one pair of trials. Because the calls were relatively long in duration, subjects were filmed for a total of 25 s following the onset of each call. Results are presented in figure 2.

Subjects responded to the playbacks either by looking in the direction of the loudspeaker or by apparently ignoring the call. In the case of hippo vocalizations, subjects generally showed little response to playback, regardless of the habitat from which the calls were played. In the case of stilt vocalizations, subjects typically responded to playback by looking in the direction of the loudspeaker, but with no significant difference in the duration of response in the dry or the wet habitat. There was some indication that the vervets recognized that the stilt's call was an alarm call: five of 18 subjects looked up and three subjects ran towards trees or stood bipedally when they heard the calls. Again, however, the monkeys did not respond more

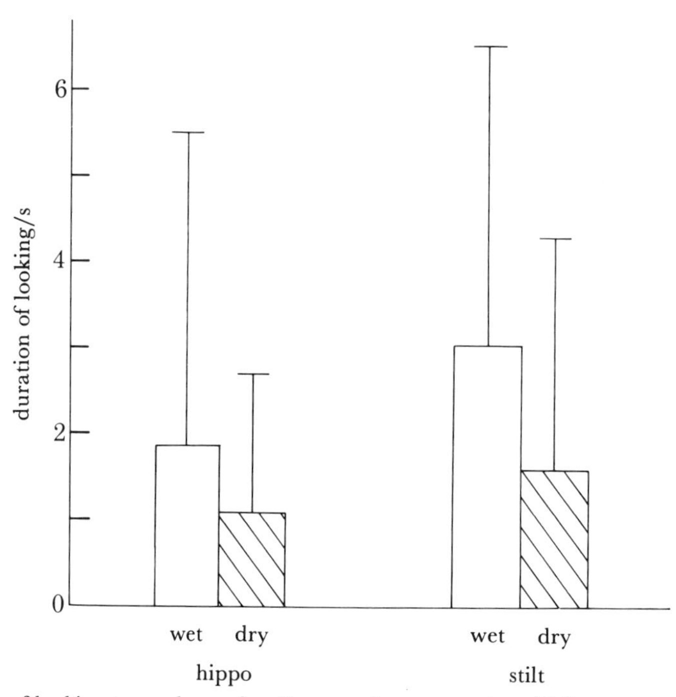

FIGURE 2. Duration of looking towards speaker (in seconds, measured at 18 frames per second) after playback of hippopotamus and black-winged stilt vocalizations from wet and dry habitats. Values shown are means and standard deviations. For hippo calls $n = 10$ subjects; for stilt calls $n = 18$ subjects. Duration of responses to calls played from different habitats did not differ significantly ($p > 0.10$).

strongly in one habitat than another. Vervets responded to both hippo and stilt calls as if they did not recognize that calls played from a dry habitat were anomalous.

These negative results, of course, cannot distinguish between the failure to recognize an anomaly and the failure to respond to one. It is entirely possible, for example, that vervets recognize that hippos belong near water, but that hippo calls played from a dry area simply fail to evoke any measurable response. Negative results *are* of interest, however, when contrasted with similar experiments that do evoke responses. Although vervets fail to respond to hippo or stilt calls coming from an inappropriate area, under comparable conditions they respond strongly to the calls of another vervet. The different performances are particularly striking given that the trials with conspecific calls asked subjects to assess the appropriate location of different *individuals*, whereas the hippo and stilt calls required only a gross understanding of the appropriate location of different *species*.

4. Associations between other species

Previous playback experiments have demonstrated that vervets can associate the screams of particular juveniles with those juveniles' mothers (Cheney & Seyfarth 1982*a*). Vervets therefore seem capable of forming associations between other group members, based on observations of their social interactions. To test whether vervets can form similar associations outside the social domain, we tested their understanding of the relationships that exist among other species.

Vervet monkeys regularly come into contact with Maasai tribesmen, who bring their cattle into the park to graze. Although the Maasai do not prey on vervets, they occasionally throw sticks or rocks at the monkeys, with the result that their approach causes increased vigilance and flight. Cows themselves pose no danger to the monkeys. Nevertheless, since cows never enter the park without Maasai, a cow alone potentially signals the approach of danger. To test whether monkeys have learned to associate cows with Maasai, we played the lowing vocalizations of either cows or wildebeest (*Connochaetes taurinus*, a common ungulate) to vervets in paired trials.

Calls were played to subjects from a mean distance of 72.3 m (s.d., 18.0). Each subject heard each type of call only once, with order of presentation varied. Because the calls were of relatively long duration, subjects were filmed for 25 s after the onset of each call type.

As figure 3 indicates, playback of cow vocalizations caused subjects to look towards the speaker for significantly longer durations than did playback of wildebeests' calls. This increased vigilance suggests that vervets associated cows with danger, and that they responded to the apparent approach of cows as they would to the approach of Maasai themselves.

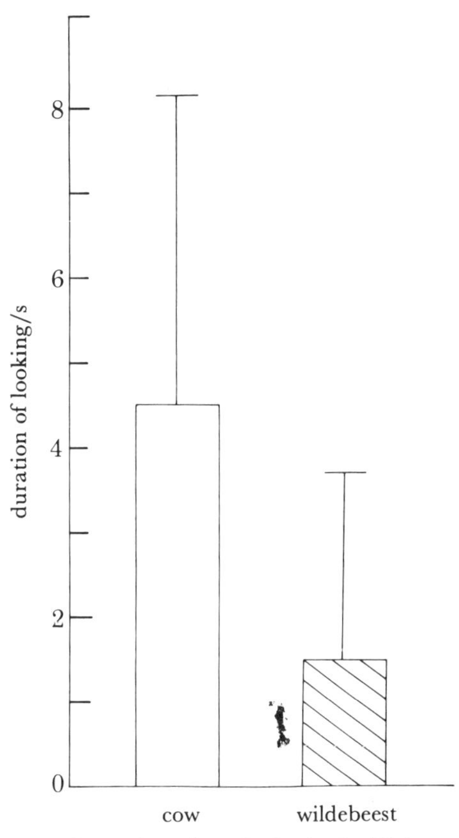

FIGURE 3. Duration of looking towards speaker after playback of wildebeest and cow vocalizations. Legend as in figure 2. Duration of responses was significantly longer after playback of cow vocalizations (two-tailed Wilcoxon test, $n = 19$, two ties, $T = 20.5$, $p < 0.05$).

5. SECONDARY CUES OF DANGER

When leopards make a kill, they frequently drag their prey into trees, where they can feed without harassment from other predators. This behaviour is peculiar to leopards, and local humans recognize that the sight of a fresh carcass in a tree denotes the proximity of a leopard. Each of the three vervet groups in our study has often seen leopards with carcasses in trees, and each time they have responded with prolonged alarm calls. We examined whether vervets knew enough about the behaviour of leopards to understand that a carcass in a tree in the *absence* of a leopard represented the same potential danger as did a leopard itself.

In conducting the experiment, we first procured a limp, stuffed carcass of a Thompson's gazelle, a species that is frequently preyed upon by leopards. This carcass was then placed in a tree, before dawn, approximately 50–75 m from the monkeys' sleeping trees. The carcass was positioned in such a way as to mimic its placement by a leopard (indeed, our attempt fooled at least one tour bus driver into thinking that a leopard was in the area). At first light, we observed the behaviour of the monkeys for a period of 2 h, noting at 5 min intervals the direction of gaze of as many group members as could be seen. In total, we presented the carcass to one group of baboons and four groups of vervets. One of these vervet groups had seen a leopard in a tree with a carcass only four days earlier, and had uttered prolonged alarm calls even when the leopard temporarily left the tree.

Despite all of the groups' experience with leopards and carcasses, neither baboons nor vervets alarm-called at the sight of the carcass alone. Moreover, there was no increased vigilance in the direction of the carcass over that which might have been expected by chance. In all cases, the monkeys behaved as if they did not recognize that a carcass in a tree denoted the proximity of a leopard.

As a further test of monkeys' knowledge of secondary cues of danger, we tested the vervets' recognition of python tracks. Pythons are a frequent predator on vervet monkeys (Cheney *et al.* 1981), and when vervets encounter a python, they give alarm calls to it and closely monitor its movements through the area (Seyfarth *et al.* 1980). Pythons lay distinct, wide, straight tracks which cannot be mistaken for those of any other species, and which are easily recognized by local humans. It is possible to determine the freshness of a python track by noting both the clarity of its outline and whether or not other species have walked across it. Indeed, on many occasions when we have encountered a fresh track we have subsequently been able to find the python in a nearby bush. Vervets in the three study groups have often watched and alarm-called at a python as it laid down a track and then disappeared into nearby bushes. Do vervets therefore recognize that a fresh python track represents potential danger?

To investigate this issue we relied upon both observation and experiments. Over a five-month period, we noted eight separate occasions when a python laid down a track in the dust and then disappeared into a nearby bush when there were no monkeys in the area. We then waited until the monkeys approached the area, and recorded their behaviour. In no case did any individual show vigilance or change its behaviour when it approached and crossed the track. Indeed, on two occasions at least one individual subsequently entered the bush where the track led, encountered the snake, and alarm-called at it. Five subsequent replications of these conditions were made by laying down an artificial python track in an area which the monkeys were approaching. Again, the animals showed no increased vigilance towards the track, and behaved as if they did not recognize that the track signalled danger.

In the preceding experiments, vervets performed well when the secondary cues of danger were auditory stimuli like alarm calls, but performed poorly when the secondary cues were visual stimuli like carcasses or tracks. There are at least two explanations for these results. First, auditory cues may be more salient than visual ones. Auditory signals have a more rapid onset time, and it has been shown that rats are more likely to associate sudden events with other sudden events, and gradual events with other gradual events (Testa 1974). Thus it may be easier for vervets to associate (and respond to) secondary cues of imminent danger when these cues are in the auditory modality. This explanation is limited, however, because it fails to explain why, in the first instance, natural selection has favoured different abilities in the visual and auditory domains.

Alternatively, it may be argued that the vervets' use of communication has evolved mainly to solve social problems, and that this has both shaped and limited their use of signals outside the social domain. Consider, for example, differences in the way primates use auditory and visual signals during social interactions. Vervets use vocal signals both in the presence and absence of visual contact. If animals are foraging in dense bush, a vocalization can tell them that another group is approaching, or that a snake has been seen nearby, without any supporting visual information (Cheney & Seyfarth 1982*b*).

In contrast, although vervets make extensive use of visual signals when communicating with each other, such signals are limited to occasions when animals are in sight of one another. Vervets do not, for example, make use of each other' tracks when foraging or monitoring incursions by neighbouring groups, nor do they visually mark aspects of their physical environment to denote their rank or group membership. As a result, their lack of attentiveness towards the visual cues of predators may be related to their limited use of visual signals as secondary cues in their social interactions. Conversely, the monkeys' regular use of auditory signals to designate objects and events may facilitate their use of auditory signals as representational cues when dealing with other species.

6. Cooperation and reciprocity

Cooperative alliances among humans are characterized by the exchange of goods or services between individuals. Significantly, such exchange is not limited to any particular domain. Exchange may involve actions (for example, reciprocal support in an aggressive coalition), individuals (the exchange of spouses between two villages), or material goods (the donation of money or food to cement an agreement). In contrast, while non-human primates frequently reciprocate past affinitive acts with future cooperation, the exchange of objects is rare (Chance 1961; Chance & Jolly 1970; Kummer 1971; Reynolds 1981).

A variety of studies has demonstrated that, in interactions involving both kin and non-kin, monkeys and apes may exchange grooming, alliances, and tolerance at food sites (for example, Packer 1977; Seyfarth 1977; Chapais & Schulman 1980; de Waal 1977, 1982). Primates seem both to remember past interactions and to adjust their cooperative acts depending on who has previously behaved affinitively towards them (Seyfarth & Cheney 1984). While monkeys and apes often reciprocate previous affinitive acts, however, such altruism rarely involves the use or exchange of objects. Primate tool use, which has received considerable attention because of its relevance to human evolution (for example, Beck 1974), is striking in part because it is relatively rare. By comparison, observers of primates are continually struck by their extraordinary ability to use other individuals as 'social tools' to achieve a particular result (for example, Kummer 1968; Chance & Jolly 1970). Similarly, although parties of baboons and chimpanzees often hunt and kill prey, there is little evidence that such hunts are truly cooperative, or that meat is genuinely shared (Kummer 1968; Altmann & Altmann 1970; Wrangham 1975; Busse 1978; Strum 1981; Teleki 1981).

Reciprocity among monkeys and apes therefore appears to occur more commonly in the form of social interactions, such as grooming and alliances, than in the exchange of material goods (Chance 1961; Reynolds 1981). Before we conclude, however, that non-human primates differ from humans in restricting their cooperative acts mainly to the social domain, a number of caveats should be mentioned.

First, the relative rarity of food sharing among non-human primates may result at least partly from the fact that, with the exception of meat, the food of non-human primates is simply not worth sharing. Monkeys and apes feed mainly on leaves and fruit that are distributed in such a way that they are not easily monopolized by one individual. There may therefore be little benefit in acquiring food directly from another. Individuals may derive greater benefit through tolerance at a particular feeding site or fruiting tree, and indeed, grooming, copulation, and other affinitive behaviour do occasionally increase the frequency with which subordinate individuals are able to feed near dominant animals (Weisbard & Goy 1976). Second, while non-human primates seldom exchange material goods for future beneficial acts, such patterns of exchange do occur in other species. For example, in the courtship displays of many birds and insects, the male offers food to his mate (reviewed in Wittenberger 1981). Finally, while non-human primates do not provision each other with food, a number of species of carnivores bring food to a central den or gathering point, where young and other individuals are fed (reviewed in Wittenberger 1981).

Cooperative behaviour in some animal species, therefore, is occasionally characterized by the exchange of material goods. We do not know, however, whether such patterns of exchange are at all modifiable. While humans can readily substitute a behavioural altruistic act for a material one, such flexibility in the 'currency' of reciprocal acts has seldom been convincingly documented in other animals. More research is clearly needed before cooperation and reciprocity in non-human species are fully understood. For the moment, however, we may hypothesize that, as in other aspects of their behaviour, reciprocity in monkeys and apes appears to occur more often in the social than in the non-social domain.

7. Discussion

When interacting with each other, vervet monkeys are apparently able to form complex associations between individuals. Within a local population, vervets can both recognize individuals and associate them with particular groups. Within their own groups, the monkeys appear to understand dominance and matrilineal kinship relations, and also remember who has behaved affinitively toward them in the past.

Vervets seem less able, however, to form similar associations about non-social aspects of their environment, even when to do so would confer an obvious selective advantage. Although the monkeys do recognize and respond to the different predator alarm calls given by birds, they appear to ignore the visual or behavioural cues associated with some predators. They do not seem to recognize the relation between a python and its track, nor do they understand that a carcass in a tree indicates a leopard's proximity, even though they have had ample opportunities to learn such associations.

Similarly, although vervets and other primates exhibit many forms of cooperation and reciprocity in their social interactions, comparable behaviour using non-social currency (for example, food sharing) is relatively rare. Monkeys readily behave altruistically and form alliances to achieve social goals, but they seldom cooperate to find or exploit new types of food resources.

Finally, vervet monkeys are poor naturalists. They seem disinclined to collect information about their environment when that information is not directly relevant to their own survival. Vervets do not seem to know that hippos stay in water during the daytime, or that particular shorebirds do not occur in dry woodlands. These data are perhaps not surprising, but they do point out a potential difference between monkeys and human beings, who are naturally curious about much of their environment, and who engage in many activities that have little practical value to survival.

We believe that these results can help us to understand the intelligence of non-human primates, and to specify more precisely how the minds of monkeys and apes differ from our own. We also recognize that any interpretation is likely to be controversial. In the following section we therefore state our own hypothesis in its strongest form, then consider alternative explanations.

The primacy of social knowledge

It is now widely agreed that species-specific predispositions affect animal learning (Seligman & Hager 1972; Hinde & Stevenson-Hinde 1973; Johnston 1981). As a result of evolution in different habitats, the behaviour of different species depends not only on the logical structure of the problems they face, but also on the particular stimuli involved. We suggest that, among primates, evolution has acted with particular force in the social domain. As a result, while monkeys are able to form and make use of complex associations in their social interactions, the same sorts of associations are formed less readily when dealing with other species. Within the social group, the behaviour of monkeys suggests an understanding of causality, transitive inference, and the notion of reciprocity. Despite frequent opportunity and often strong selective pressure, however, comparable behaviour does not readily emerge in dealings with other animal species or with inanimate objects.

The special sensitivity of non-human primates to social events is not surprising. Human infants, after all, show special sensitivity to social as opposed to non-social visual stimuli (Sherrod 1981), to speech sounds as opposed to other auditory stimuli (Eimas *et al.* 1971), and to human interactions as opposed to other relations in their environment (see below). In a similar manner, non-human primates appear to exhibit their most subtle discriminations when dealing with conspecific faces, sounds, and social relations.

Because primate intelligence has evolved mainly to solve social problems, monkeys often show surprising gaps in their knowledge of the non-social world. For example, in social interactions visual cues are not used to represent objects or individuals in the absence of face-to-face encounters. Auditory cues, in contrast, often function during social behaviour to designate objects in the absence of visual information. These differences are reflected in the non-social domain, where monkeys are 'prepared' (Seligman 1970) to recognize auditory cues that are secondary indicators of predators, but appear 'unprepared' to associate visual cues with danger. Differences in the monkeys' use of visual and auditory information when dealing with other species may therefore result from the different way they communicate in these modalities when dealing with their own species.

Alternative arguments

(*a*) *Differences between apes and monkeys*

Some of our generalizations about domain-specific performance may be less applicable to apes than to monkeys (Premack 1976), since apes do appear to make occasional use of visual symbols in their social interactions. Free-ranging chimpanzees, for example, make sleeping nests each night, and, when the members of one group make incursions into the range of another, they have been observed to make aggressive displays upon encountering their neighbours' empty nests (Goodall *et al.* 1979). The captive chimpanzee Vicki was able to sort pictures of animate and inanimate objects into distinct categories without previous training (Hayes & Nissen 1971). Whether or not a monkey would be capable of similar classification is not known, because the relevant experiments have not yet been conducted.

(*b*) *The importance of ecological factors*

The food exploited by non-human primates, particularly ripe fruit, is both spatially and temporally dispersed. Field data on many species, especially orangutans and chimpanzees, indicate that primates frequently range over large areas, and that they remember the locations and phenological patterns of both water and a variety of plant foods (Clutton-Brock 1977; Rodman 1977; Wrangham 1977; Sigg 1980; Sigg & Stolba 1981). As a result, it has been argued that ecological pressures have played a major role in the evolution of primate intelligence (for example, Clutton-Brock & Harvey 1980; Milton 1981).

This hypothesis emphasizes that the distinction we have drawn between social and non-social knowledge is not a simple one. Primate memory has no doubt evolved as a result of the need to remember both the location of spatially dispersed food resources and previous social encounters. The point is not to oppose one unifactorial ecological argument against an equally unifactorial social one, but to gain a better understanding of precisely how ecological and social factors have combined to give non-human primates an intelligence that appears simultaneously to be superior to that of other mammals and inferior to our own.

Although ecological factors are undoubtedly important, primates do not appear to manipulate objects in their environment to solve ecological problems with as much sophistication as they manipulate each other to solve social problems (see above). The challenge of exploiting widely dispersed and ephemeral food items may thus have led to increased intelligence not simply because food collection itself becomes more difficult, but also because ecological complexity sets the stage for increasingly complex social competition.

Social knowledge and non-social knowledge in human infants

In the past, many students of human development believed that infants' knowledge of the social and non-social world developed at similar rates (for example, Piaget 1963). Recent studies question this view, and suggest that an understanding of certain concepts may appear at an earlier age when the stimuli involved are animate (especially other people) than when they are inanimate. For example, Hood & Bloom (1978) examined the development of children's expressions of causality, using two- and three-year-olds as subjects. Previous work had indicated that causal understanding develops slowly, and is often not apparent until age seven or eight (Piaget 1963). Hood & Bloom, in contrast, found that children readily discussed the intentions and motivations of people in causal terms. They did not, however, talk about causal events involving objects (see also references in Gelman & Spelke 1981; Hoffmann 1981). In a study of naming behaviour in children 17–22 months old, MacNamara (1982, p. 30) concluded: '...by the time the child comes to learn language, he has already learned that objects in certain categories are important as individuals, those in other categories are merely exemplars of the category. Person is the preeminent category of the first sort'.

These results, together with those presented above, suggest that there is, in both human and non-human primates, an evolutionary predisposition which makes it easier for organisms to understand relations among conspecifics than to understand similar relations among things. Compared with humans, non-human primates exhibit this predisposition in an extreme form: they show sophisticated cognitive skills when dealing with each other, but exhibit such skills less readily in their interactions with objects. Among humans the predisposition is more subtle, but nevertheless may appear in the earliest years of childhood, when infants exhibit remarkable social skills while at the same time remaining ignorant of much of the world around them. For a few brief years, children reveal the results of selection acting on the primate brain: selection that has made them particularly sensitive to the emotions, behaviour and social relations of their conspecifics.

Research supported by N.S.F. grant BNS 82-15039. We thank the Office of the President, Republic of Kenya, for permission to conduct research in Amboseli National Park, and Bob R. O. Oguya, the Warden in Amboseli, for help and cooperation. We thank Bernard Musyoka Nzuma for field assistance, E. Mottram and the staff of the National Museums of Kenya for help in preparing a Thompson's gazelle carcass, and D. and C. Nightingale for help in recording cow vocalizations. We thank D. Dennett, S. Essock-Vitale, D. R. Griffin, H. Kummer, C. Ristau, B. Smuts and F. de Waal for their comments on earlier drafts. During manuscript preparation the authors were fellows at the Center for Advanced Study in the Behavioral Sciences, where they received support from N.S.F. (BNS 76-22943) and the Sloan Foundation (82-2-10).

References

Altmann, S. A. & Altmann, J. 1970 *Baboon ecology*. Chicago: University of Chicago Press.

Bachmann, C. & Kummer, H. 1980 Male assessment of female choice in hamadryas baboons. *Behav. Ecol. Sociobiol.* **6**, 315–321.

Beck, B. B. 1974 Baboons, chimpanzees, and tools. *J. Hum. Evol.* **3**, 509–516.

Busse, C. D. 1978 Do chimpanzees hunt cooperatively? *Am. Nat.* **112**, 767–770.

Chance, M. R. A. 1961 The nature and special features of the instinctive social bond of primates. *Viking Fund Publ. Anthrop.* **31**, 17–33.

Chance, M. R. A. & Jolly, C. 1970 *Social groups of monkeys, apes, and men*. London: Jonathan Cape.

Chapais, B. & Schulman, S. 1980 An evolutionary model of female dominance relations in primates. *J. theor. Biol.* **82**, 47–89.

Cheney, D. L. 1978 Interactions of immature male and female baboons with adult females. *Anim. Behav.* **26**, 389–408.

Cheney, D. L. 1983 Extra-familial alliances among vervet monkeys. In *Primate social relationships* (ed. R. A. Hinde), pp. 278–286. Oxford: Blackwell.

Cheney, D. L., Lee, P. C. & Seyfarth, R. M. 1981 Behavioral correlates of non-random mortality among free-ranging adult female vervet monkeys. *Behav. Ecol. Sociobiol.* **9**, 153–161.

Cheney, D. L. & Seyfarth, R. M. 1980 Vocal recognition in free-ranging vervet monkeys. *Anim. Behav.* **28**, 362–367.

Cheney, D. L. & Seyfarth, R. M. 1981 Selective forces affecting the predator alarm calls of vervet monkeys. *Behaviour* **76**, 25–61.

Cheney, D. L. & Seyfarth, R. M. 1982*a* Recognition of individuals within and between groups of free-ranging vervet monkeys. *Am. Zool.* **22**, 519–529.

Cheney, D. L. & Seyfarth, R. M. 1982*b* How vervet monkeys perceive their grunts. *Anim. Behav.* **30**, 739–751.

Cheney, D. L. & Seyfarth, R. M. 1983 Non-random dispersal in free-ranging vervet monkeys. *Am. Nat.* **122**, 392–412.

Cheney, D. L. & Seyfarth, R. M. 1984 The manipulation of alarm calls by vervet monkeys. *Behaviour* (In the press.)

Cheney, D. L., Seyfarth, R. M., Andelman, S. J. & Lee, P. C. 1985 Factors affecting reproductive success among vervet monkeys. In *Reproductive success* (ed. T. H. Clutton-Brock). Chicago: University of Chicago Press.

Clutton-Brock, T. H. (ed.) 1977 *Primate ecology*. New York: Academic Press.

Clutton-Brock, T. H. & Harvey, P. 1980 Primates, brains, and ecology. *J. Zool., Lond.* **190**, 309–323.

Datta, S. B. 1983 Relative power and the acquisition of rank. In *Primate social relationships* (ed. R. A. Hinde), pp. 93–103. Oxford: Blackwell.

Dohl, J. 1968 Über die Fahigkeit einer Schimpansin, umwege mit selbstandigen zwischenzielen zu uberblicken. *Z. Tierpsychol.* **25**, 89–103.

Eimas, P., Siqueland, P., Jusczyk, P. & Vigorito, J. 1971 Speech perception in infants. *Science, Wash.* **171**, 303–306.

Gelman, R. & Spelke, E. 1981 The development of thoughts about animate and inanimate objects. In *Social cognitive development* (ed. J. H. Flavell & L. Ross), pp. 43–66. Cambridge: Cambridge University Press.

Gillan, D. J. 1981 Reasoning in the chimpanzee. II. Transitive inference. *J. exp. Psychol. Anim. Behav. Proc.* **7**, 150–164.

Goodall, J., Bandoro, A., Bergmann, E., Busse, C., Matama, H., Mpongo, E., Pierce, A. & Riss, D. 1979 Intercommunity interactions in the chimpanzee population of the Gombe National Park. In *The great apes* (ed. D. Hamburg & E. R. McCown), pp. 13–54. Menlo Park: Benjamin Cummings.

Gouzoules, S., Gouzoules, H. & Marler, P. 1984 Rhesus monkey screams: representational signalling in the recruitment of agonistic aid. *Anim. Behav.* **32**, 182–193.

Hayes, K. J. & Nissen, C. H. 1971 Higher mental functions of a home-raised chimpanzee. In *Behavior of non-human primates* (ed. A. M. Schrier & F. Stollnitz), pp. 60–115. New York: Academic Press.

Hinde, R. A. & Stevenson-Hinde, J. 1973 *Constraints on learning*. New York: Academic Press.

Hoffman, M. L. 1981 Perspectives on the difference between understanding people and understanding things: the role of affect. In *Social cognitive development* (ed. J. H. Flavell & L. Ross), pp. 67–81. Cambridge: Cambridge University Press.

Hood, L. & Bloom, L. 1979 What, when, and how about why. *Monogr. Soc. Res. Child Dev.* **44**, 1–41.

Humphrey, N. K. 1976 The social function of intellect. In *Growing points in ethology* (ed. P. Bateson & R. A. Hinde), pp. 303–318. Cambridge: Cambridge University Press.

Johnston, T. D. 1981 Contrasting approaches to a theory of learning. *Behav. Brain Sci.* **4**, 125–173.

Jolly, A. 1966 Lemur social behavior and primate intelligence. *Science, Wash.* **153**, 501–506.

Judge, P. G. 1982 Redirection of aggression based on kinship in a captive group of pigtail macaques. *Int. J. Primatol.* **3**, 301.

Kummer, H. 1968 Social organization of hamadryas baboons. Chicago: University of Chicago Press.

Kummer, H. 1971 *Primate societies*. Chicago: Aldine Publishing Co.

Kummer, H. 1982 Social knowledge in free-ranging primates. In *Animal mind – human mind* (ed. D. R. Griffin), pp. 113–130. New York: Springer-Verlag.

MacNamara, J. 1982 *Names for things*. Cambridge, Massachusetts: M.I.T. Press.

McGonigle, B. O. & Chalmers, M. 1977 Are monkeys logical? *Nature, Lond.* **267**, 694–696.

Milton, K. 1981 Distribution patterns of tropical plant food as an evolutionary stimulus to primate mental development. *Am. Anthrop.* **83**, 534–548.

Olivier, R. C. D. & Laurie, A. 1974 Habitat utilization by hippopotamus in the Mara River. *E. Afr. Wildl. J.* **12**, 249–271.

Packer, C. 1977 Reciprocal altruism in *Papio anubis*. *Nature, Lond.* **265**, 441–443.

Piaget, J. 1963 *The psychology of intelligence*. New York: International Universities Press.

Premack, D. 1976 *Intelligence in ape and man*. Hillsdale: Lawrence Erlbaum.

Premack, D. & Premack, A. 1982 *The mind of an ape*. New York: Norton.

Reynolds, P. C. 1981 On the evolution of human behavior. Berkeley: University of California Press.

Rodman, P. S. 1977 Feeding behavior of orangutans in the Kutai Nature Reserve, East Kalimantan. In *Primate ecology* (ed. T. H. Clutton-Brock), pp. 384–414. New York: Academic Press.

Rozin, P. 1976 The evolution of intelligence and access to the cognitive unconscious. In *Progress in psychology*, vol. 6 (ed. J. N. Sprague & A. N. Epstein), pp. 245–280. New York: Academic Press.

Seligman, M. E. P. 1970 On the generality of the laws of learning. *Psych. Rev.* **77**, 406–418.

Seligman, M. E. P. & Hager, J. L. 1972 *Biological boundaries of learning*. New York: Appleton Century Crofts.

Seyfarth, R. M. 1977 A model of social grooming among adult female monkeys. *J. theor. Biol.* **65**, 671–698.

Seyfarth, R. M. 1981 Do monkeys rank each other? *Behav. Brain Sci.* **4**, 447–448.

Seyfarth, R. M. & Cheney, D. L. 1984 Grooming, alliances, and reciprocal altruism in vervet monkeys. *Nature, Lond.* **308**, 541–543.

Seyfarth, R. M., Cheney, D. L. & Marler, P. 1980 Vervet monkey alarm calls. *Anim. Behav.* **28**, 1070–1094.

Sherrod, L. R. 1981 Issues in cognitive-perceptual development: the special case of social stimuli. In *Infant social cognition* (ed. M. E. Lamb & L. R. Sherrod), pp. 11–36. Hillsdale, New Jersey: Lawrence Erlbaum Associates.

Sigg, H. 1980 Differentiation of female positions in hamadryas one-male units. *Z. Tierpsychol.* **53**, 265–302.

Sigg, H. & Stolba, A. 1981 Home range and daily march in a hamadryas baboon troop. *Folia Primatol.* **36**, 40–75.

Smuts, B. 1985 *Sex and friendship in baboons*. Chicago: Aldine Publishing Co.

Strum, S. C. 1981 Processes and products of change: baboon predatory behavior at Gilgil, Kenya. In *Omnivorous primates* (ed. R. Harding & G. Teleki), pp. 255–302. New York: Columbia University Press.

Teleki, G. 1981 The omnivorous diet and eclectic feeding habits of chimpanzees in Gombe National Park, Tanzania. In *Omnivorous primates* (ed. R. Harding & G. Teleki), pp. 303–343. New York: Columbia University Press.

Terrace, H., Pettito, L., Sanders, R. & Bever, T. 1979 Can an ape create a sentence? *Science, Wash.* **206**, 891–902.

Testa, T. J. 1974 Causal relationships and the acquisition of avoidance responses. *Psychol. Rev.* **81**, 491–505.

de Waal, F. 1977 The organization of agonistic relations within two captive groups of Java monkeys. *Z. Tierpsychol.* **44**, 225–282.

de Waal, F. 1982 *Chimpanzee politics*. New York: Harper & Row.

Weisbard, C. & Goy, R. 1976 Effect of parturition and group composition on competitive drinking order in stumptail macaques. *Folia primatol.* **25**, 95–121.

Wittenberger, J. 1981 *Animal social behavior*. Boston: Duxbury Press.

Woodruff, G. & Premack, D. 1979 Intentional communication in the chimpanzee: the development of deception. *Cognition* **7**, 333–362.

Wrangham, R. W. 1975 Behavioral ecology of chimpanzees in Gombe National Park, Tanzania, Ph.D. Thesis, University of Cambridge.

Wrangham, R. W. 1977 Feeding behavior of chimpanzees in Gombe National Park, Tanzania. In *Primate ecology* (ed. T. H. Clutton-Brock), pp. 504–538. New York: Academic Press.

Phil. Trans. R. Soc. Lond. B **308**, 203–214 (1985) [203]
Printed in Great Britain

Conditions of innovative behaviour in primates

By H. Kummer[1] and Jane Goodall[2]

[1] *Ethologie and Wildforschung, Universität Zürich-Irchel II, Winterthürerstrasse* 190, *CH*-8057 *Zürich, Switzerland*
[2] *P.O. Box* 727, *Dar-es-Salaam, Tanzania*

Innovative behaviour achieved through exploration, learning and insight heavily depends on certain motivational, social and ecological conditions of short duration. We propose that more attention should be given to what these conditions are and where they are realized in natural groups of non-human primates. Only to the extent that such favourable conditions were frequently realized in a social structure or an extraspecific environment could selective pressures act on innovative abilities. There is hope that research into field conditions of innovative behaviour will help to identify its selectors in evolution.

Part 1 (H. Kummer)

Introduction

The motive of this paper is that we almost completely lack an ecology of intelligence. No other dimension of behaviour has so systematically *not* been studied in the field.

The major selection pressures that made the order of primates into specialists for mental flexibility have been the subject of some speculation. Chance & Mead (1953) seem to have been the first to develop the argument that the complexity of *social* life was the prime selective agent of primate intelligence. The group member is required to judge the skills and changing inclinations of others and to use them for intelligent predictions for its own benefit. The view was later and independently presented by Alison Jolly (1966) and by Humphrey (1976). The difficulty is that many non-primates live in large groups without having evolved particular intelligence levels. Social complexity is probably the effect as much as the ultimate cause of intelligence.

Katharine Milton (1981) plausibly argued that the necessity of predicting the few productive days of widely dispersed tree patches in tropical forests was a major factor in the evolution of primate mental abilities. Yet, resources dispersed in time and space are hardly a unique feature of primate environments; they are also mastered by predators of highly mobile groups of prey. The complex system of ultimate causes is likely to defy any one-factor explanation. While certain selective pressures may have been particularly formative, it is most probable that a whole *set* of primate characters were mutually preadaptive to each other's evolution, and that they evolved into a syndrome of which mental flexibility is a part. Foremost among these characters is the *a priori* lack of outstanding competence in any specific skill. Our inventive species probably emerged among the primates rather than from a different order *because* primates are remarkably ill-equipped with innate technologies. The termite is better than the chimp at nest-building by an incredible amount, but the chimp beats the termite by learning to tease him from his mound. Instinct is like a key fitting a single lock. A key is easy to use. The poor *a priori* competence of primates is comparable to a pick-lock. Its use requires and therefore promotes a kind of skill which eventually permits the opening of many locks.

If some selectors of primate intelligence were of outstanding importance, it should be possible to locate them empirically. The rationale is this: a behavioural character is tested by selection only in certain situations, those in which the behaviour is used to function. The critical assumption is that the situation that *selected* for a behavioural character in phylogeny also *trains* it in individual life. For example, the genetic basis of flying skill was selected only while the primeval birds were airborne; the learning of flying skill is restricted to the same context. An intelligence or learning disposition selected by the context of being socially subordinate would presumably be most proficiently developed by subordinates. The extent to which this assumption is correct must be ascertained for each behaviour. In general, it should be correct (i) if the behaviour with the higher fitness contribution were also the one with the greater reward – a very frequent case in all animals; (ii) if each individual were capable of several behavioural variants so that it could compare the rewards – a frequent case in primates; and (iii) if training could substantially improve on the uninstructed inherited ability – a condition fulfilled by definition in learning and intelligence.

From the above one may assume that innovative behaviour was selected primarily in the situations that now most effectively teach it. It would therefore be profitable to look systematically at the everyday situations in which present-day primates learn and invent most proficiently. These are the conditions by and for which their innovative dispositions must have been selected.

Time does not permit even a sketchy classification of the everyday conditions relevant to innovative performance. Instead a few examples from the field and the quasi-natural captive group will show the interest of the topic.

Intellectual training by group life

If the Chance–Jolly–Humphrey hypothesis is correct, that is, if complex social life selected primate intelligence *and* trained it to its greatest proficiency in natural life, we would expect the training effect to be most pronounced in those group members whom group life affects and constrains most, compared with solitaries. Most likely, these are the group members of low dominance rank. Subordinates should have made the most of the genetic disposition for intelligence and learning, since they most depend on the formation of alliances or need to lead their mate to where the dominant male cannot see them, and so forth.

Strayer (1976) and Bunnell *et al.* (1980; Bunnell & Perkins 1980) tested macaques that normally lived in captive groups on learning tasks. In Strayer's study, a buzzer sounded when a reward could be obtained. The subordinate monkeys made fewer time-out errors, that is, they did not press the lever when the buzzer was silent, whereas dominants pressed even with the buzzer off. Dominant macaques were more dependent on regular reinforcement. In Bunnell's study the subordinate macaques were faster to master a reversal learning task. Dominant animals that fell in rank improved their performance, which speaks against the possibility that subordinates performed better because they were younger. The subordinates in these studies were thus more attentive to the outside conditions of their actions as the test apparatus, and they were better at relearning and at persisting in spite of frustration. This is what we would expect if they had transferred abilities acquired in the highly conditional social life of a low-ranker.

Free time and energy

These are other candidate conditions of innovative behaviour. The reason is that environments are not stationary over time. Animals survive certain periods in which their maximized net energy intake is just sufficient to keep them alive. At other more favourable periods, survival may be so easy that the animal has time and energy to spare. It would be advantageous if it could somehow store them for worse times.

To hoard food in a cache is one form of storage, but the hoard is vulnerable to decay, to parasitism and to plundering by observant conspecifics. It is found in solitary rodents and in insects with hidden or defended nests. Non-human primates do not use hoards, probably because they have no dens and no organization for guarding and sharing. A second form of storage is fatty tissue, which is a well-defendable but burdensome hoard. It is particularly common in marine mammals, where part of the weight is carried by the water and where the fat is a useful insulator against hypothermia. Primates, being arboreal and a prey species, require agility and hardly use it.

By far the most elegant form of storage is new knowledge or skill in communication or ecological techniques. Here, the time and energy spent in exploratory behaviour is transformed into a light-weight unstealable commodity that can save time and energy in a harsher future.

Every zoo colony of primates is a rough experimental test of the hypothesis that free time and energy do indeed promote innovation. Kummer & Kurt (1965) quantitatively compared the behaviour of hamadryas baboons in the Zurich Zoo with the behaviour of wild troops in Ethiopia. Of 68 motor and vocal signals that the zoo baboons used in social communication, nine were not found in any wild troop, suggesting that they were innovations. (On the other hand, all social signals seen in the wild were also found in the zoo colony.) None of them looked pathological. Rather, they were functional elaborations, such as a new gesture that invited a youngster to be carried. In the case of 'protected threat', in which a female directs the attack of the dominant male against her opponent, the zoo version was technically improved and clearly more efficient than the crude hints at the same behaviour in the wild.

The best ecological invention of the zoo baboons was to obtain drinking water from a mound, otherwise unobtainable, by lowering the body, tail first, down a vertical wall and then to suck the water from the tuft (Schönholzer 1958). The tail-drinking also was an innovation. No wild baboon was seen to approach an incentive backwards.

Within the limits, free time and energy thus seem to further innovation.

Part 2. (Jane Goodall)

Introduction

The study of the chimpanzees at Gombe, in Tanzania, is now in its 24th year. The longitudinal records provide many anecdotes regarding innovative behaviour and the following discussion concerning conditions favouring (i) the appearance of novel patterns *per se*; and (ii) the transmission of such patterns through the social group, is centred around this information.

An innovation can be: a solution to a novel problem, or a novel solution to an old one; a communication signal not observed in other individuals in the group (at least at that time) or an existing signal used for a new purpose; a new ecological discovery such as a food item not previously part of the diet of the group. In some species of primates a new food becomes incorporated into the diet of a troop as a result of a whole series of independent individual discoveries; other novel behaviours, such as potato washing, are discovered by one individual only and acquired by other troop members through observational learning.

Some innovations derive from the ability of the individual to profit from an accidental happening. Thus when the chimpanzee Mike was afraid to take a banana from my hand he seized a clump of grasses and swayed them to and fro in the typical threat gesture. As he did so one of the grasses touched the fruit. Instantly he let go of the grasses, looked round, hastily broke off a slender, bendy twig, dropped it at once, reached for a thicker stick and hit the banana to the ground from where he took and ate it. Ten minutes later when I proffered a second fruit Mike unhesitatingly reached for a stick and repeated the performance. The first step in such a chance sequence is not always seen, but can sometimes be deduced. At Gombe, pots and pans are often washed at the edge of the lake. The anubis baboons of one study troop spend long periods of time digging in the gravel, searching for scraps. One day when the lake was rough and the successive breaking of large waves made drinking difficult, an adolescent male, Asparagus, was observed to dig a hole near the high water mark, wait for a wave to fill it, then quickly drink as the wave retreated. Subsequently he was observed to do the same thing on another occasion. Six years later eight-year-old Sage, of the same troop, was seen drinking in the same way on four separate occasions. It seems likely that the two males discovered the technique independently – probably by first drinking water that filled up holes dug by them when searching for food and subsequently profiting from that experience, digging the holes for the specific purpose of drinking. One year after he was seen drinking in this way Asparagus disappeared (he died or transferred to a new troop). Sage was three years old at that time, and might, therefore, have learnt the behaviour by watching the older male: if so it seems strange that he was not seen to drink in this way during the next six years.

Other innovations result from the ability of the higher primates to use existing behaviour patterns for new purposes. Chimpanzees at Gombe routinely use leaves to wipe their bodies when soiled: one female twice used handfuls of leaves to brush away stinging insects (once bees from a hive she was raiding, once ants from the trunk of a tree to which she was clinging as she raided their nest) and one male used leaves to wipe the inside of a baboon skull during a meat-eating session: he ate the leaf-brain wad (Wrangham 1975).

A third kind of innovation is the performance of a completely new pattern, such as the bizarre series of somersaults by which the polio stricken male, McGregor, managed to move from place to place after losing the use of his legs or the odd form of locomotion invented by one infant during lone play (to be described).

Conditions for the appearance of new behaviours

Innovations may be occasioned by sudden change in the environment such as when wild populations are provisioned or during an outbreak of disease (such as the polio at Gombe (Goodall 1983)). They may also, as has been described, appear during periods of environmental stability, particularly during periods of plenty which result in an excess of leisure and energy. This is exemplified in some captive situations.

A prerequisite for the solving of problems in novel ways, be they ecological, technical (involving the use of tools) or social (involving the manipulations of companions) is familiarity with the components of the situation. Chimpanzees have certain manipulative tendencies, which appear to be inborn – at a certain age an infant will reach out, grasp and brandish a stick (Schiller 1949). But he must become familiar with sticks during play if he is to use them successfully as tools to solve novel problems (figure 1) (Köhler 1925; Schiller 1952). Similarly, chimpanzees have certain inborn communications signals, but must have opportunity to interact with other chimpanzees to use these signals successfully (Menzel 1964). Social experience and familiarity with the other members of the group are essential if a chimpanzee (or other animal) is to invent novel methods of getting his own way in competitive situations.

FIGURE 1. Flint 'clubs' an insect, a behaviour not observed in the adult tool-using repertoire at Gombe.

(a) Social innovations

Much social skill emerges as a direct result of competition between group members. A subordinate individual A, to attain a particular goal *in spite of* the inhibiting proximity of a social superior B must either form a coalition with a third individual C (so that A and C are superior in strength to B) or follow a more devious route, such as moving out of B's sight, or persuading B to move away, or distracting B's attention. Some of the solutions to social problems of this sort represent novel ones on the part of the individual concerned. The following are examples of innovation of this sort observed among the chimpanzees at Gombe.

(i) One reproductive strategy of the male chimpanzee is to establish exclusive mating rights over a female by leading her away from the other males to the periphery of the community range (McGinnis 1973; Tutin 1979). A male wanting to establish a consortship of this sort faces obvious problems if there are higher ranking males in the group who are interested in the same female. One individual, Satan, was, on several occasions observed to keep very close to a sexually attractive female until she made her nest in the evening, rise before any other chimpanzee the next morning, shake branches at her (a signal for her to follow him) and lead her away before the others left their nests (Tutin 1979).

(ii) The charging display, when a male rushes over the ground or through the branches, swaying, dragging and throwing vegetation, branches and rocks is of major significance to his struggle for a higher position in the dominance hierarchy. Figan not only showed unusual timing and placing of his displays (Bygott 1974) but repeatedly got up before other chimpanzees in his group and performed wild arboreal displays while his companions were still in their nests. This caused great confusion and was one of the methods he employed during his successful bid for alpha position (Riss & Goodall 1977). Other males occasionally used this technique, but it seems that only Figan profited from the experience and made it a regular occurrence.

(iii) Adolescent males typically challenge older females with bristling, swaggering displays during which branches may be waved with which they may hit the females. Responses to such displays vary depending on the age of the male, rank of his victim in relation to that of his mother and so on, and range from fear and avoidance, through retaliation, to totally ignoring the youngster for as long as possible. Sometimes they hold onto the branches with which they are being flailed. One female, Winkle, twice terminated such displays in unusual ways. Once she firmly removed the branch from a young male even before he had begun to sway it, and once she reached out and vigorously tickled the swaggering youngster so that his aggressive display ended in *laughing*. Of interest was the fact that the same female used two quite different unusual patterns to achieve the same end.

Banana-feeding at Gombe created a new situation. Aggressive competition was greatly increased and the chimpanzees tended to aggregate in large numbers in camp as they waited for us to distribute the fruits. Among the innovations that appeared as a result of the new situation were the following.

(i) Before we had devised a way of distributing the bananas among the various individuals present it was often the case that youngsters got very few or none at all. One day, when the young adolescent male Figan was part of a large group and, in consequence, had not managed to get more than a couple of bananas, he suddenly got up and walked away with a purposeful gait. His mother followed and, as is often the case when one individual sets off as though with a goal in mind, the others followed. Ten minutes later Figan reappeared by himself, and was given bananas. We thought this was a coincidence as indeed it may have been on that first occasion. But when Figan repeated the performance on four subsequent occasions it became clear that he was following a deliberate strategy (Goodall 1971).

(ii) In the early sixties there were occasionally a few empty four gallon paraffin cans lying about my camp. At one time or another almost all of the 14 adult males had incorporated one of these into a charging display. One of them, Mike, profited from the experience and began to use the cans in almost all displays that he performed in camp. He even learned to keep three ahead of him as he charged towards his superiors. Within a four-month period he had become alpha male, having thoroughly intimidated all his rivals and, so far as we know, without taking part in a single fight (Goodall 1968, 1971).

Groups of chimpanzees in captivity show, as did the hamadryas, much innovation in the social sphere. With adequate food and drink provided, their health attended to, and most dangers and excitements removed from daily life, chimpanzees not only have more leisure time than their wild counterparts, they are also subjected to a new stress, over and above boredom and confinement. At Gombe, if the social environment becomes tense and hostile, an individual can, and often does, move off quietly by itself or with a small compatible group. In captivity, even in a large enclosure, this is not possible. The novel behaviours which zoo or laboratory chimpanzees show in their social interactions are undoubtedly, at least in part, a response to this new challenge. And, because they are forced to remain constantly in each other's company, chimpanzees in captivity are probably able to predict more accurately the responses of their companions so that their interactions can become increasingly sophisticated. One adolescent male, Shadow, in Emil Menzel's group, developed a unique courtship display during which he stood bipedally and flipped his upper lip back over his nostrils. Precisely how this strange performance originated is not known: it was clear, however, that the adult females, all dominant to Shadow, responded aggressively to the more usual male courtship (probably because this involves many aggressive elements). With the new display Shadow was able to convey his sexual interest devoid of aggressive overtones (C. Tutin & W. McGrew, personal communication). The chimpanzees of the large Arnhem colony have been carefully studied by de Waal (1982) who observed a great deal of extraordinarily astute social skill. Once, for example, when a subordinate male was surprised during a clandestine courtship by the arrival of the alpha male, he quickly covered his erect penis with his hands, effectively hiding the tell-tale signal. After an aggressive dispute between two of the adult males, a female sometimes showed behaviour which served to hasten reconciliation between the rivals: she would lead one towards the other, sit between them so that both groomed her, then step quietly away leaving the males grooming each other: harmony was restored. Each of the adult females was seen to act thus at one time or another; such incidents have never been observed in the wild.

(*b*) *Ecological and technical innovations*

While social innovations can only appear within the social environment, during interactions with other group members, ecological and technical innovations are more likely to appear during periods when an individual is, to some extent, freed from social distractions. In a hamadryas one-male unit the females have differing spatial positions: one stays close to the unit male; one spends much time at the periphery of a foraging group. While the central female is preoccupied with social interactions, particularly with the male, the peripheral female is alert to danger and is the individual most likely to discover a new food source. Sigg (1980) tested wild family units temporarily kept in field enclosures: he found that there was a marked difference in the abilities of peripheral versus central females in their ability to learn new ecological tasks. Thus peripheral females readily learned to discriminate between nails painted different colours that were used as markers for food buried in the ground. They also remembered the location of underground water several hours after being allowed to watch its burial. Central females seemed unable to learn these tasks, perhaps because their attention was almost totally absorbed in social interactions. In the Tai Forest of the Ivory Coast chimpanzees crack open nuts with stones or heavy pieces of wood. Adult females are more efficient in the hammering technique than males on several measures. This may be due, at least partly, to the fact that males appear to be more easily distracted by ongoing social events: they look around, away from the task in hand, more frequently than do hammering females (Boesch & Boesch 1981, 1983).

Many innovations appear during childhood when a youngster is cared for and protected by its mother and thus has much time for carefree play and exploration. At Gombe, where female chimpanzees spend a great deal of time in association with her dependent offspring only, an infant may be without opportunity for playing with other youngsters for hours or even days at a stretch. With no social distraction the conditions are ideal for the emergence of novel behaviours in the ecological and technical sphere. This is particularly true for the first-born child who does not even have the opportunity to play with elder siblings. One infant invented a new method of locomotion, swinging her body forward through her arms (in the typical 'crutching' gait) while keeping one foot firmly tucked into the opposite groin. Another threw a round fruit into the air and caught it as it fell: he tried to repeat the performance many times but I did not see him succeed again. A third lay on his back with his legs in the air and rotated a large round fruit which he held on the soles of his feet. A fourth used stones and twigs to tickle her clitoris, continuing until she was laughing loudly. One infant spent many minutes playing with a butterfly, another with a frog. Occasionally an infant will taste a novel food object: behaviour which we have never observed in an adult chimpanzee in the wild.

Once an infant has mastered a newly acquired adult manipulative pattern this may be practised in a variety of contexts other than that in which it was learned. Thus one infant, Flint, used the termite-fishing technique on his mother's leg, pushing a blade of grass between the hairs and then sucking the end. And, on another occasion, Flint used a grass blade to 'fish' for water caught in the hollow of a tree stump. After sucking off the drops a few times the blade gradually became more and more crumpled until he had made a miniature *sponge* (Figure 2). This

FIGURE 2. (*a*) Flint dips a grass stem into a water bowl, using the termite-fishing technique. (*b*) After a few such dips the grass becomes crumpled and resembles a miniature *sponge*. (A sponge of crumpled leaves is traditionally used by the Gombe chimpanzees for drinking from water bowls.)

behaviour might have led to the typical leaf sponge used by the Gombe chimpanzees for soaking water from hollows of this sort (although it seems more likely that this originated for the removal of dead leaves that had accumulated in the water). At Gombe the chimpanzees open the large hard-shelled *Strychnos* fruits by banging them against the tree trunk or a rock. During one *Strychnos* season infant Flint picked up a rock and smashed an insect on the ground. This pattern might, one day, lead to the use of a hammer stone for cracking open nuts, as in West African chimpanzee cultures. On another occasion Flint hit an insect with a 'club' (figure 3).

FIGURE 3. (*a*) Feeding traditions are passed from one generation to the next through mechanisms of social facilitation and observational learning. Here one-and-a-half-year-old Getty watches his grandmother's six-year-old son (his uncle), Gimble, feeding on leaves. (*b*) Getty then samples the same food, while Gimble, in turn watches him. This was not a new item in the diet, but the example illustrates the mechanisms for the passing on of information from one individual to another in chimpanzee society.

(*c*) *Innovation and tradition*

A comparison of the ecological, technical and social communication patterns of different groups of chimpanzees in different geographical localities reveals a wide variety of cultural traditions. Gombe and Mahale, 100 miles (160 km) apart, have many plant species in common: Mahale chimpanzees feed on some that are never used by the Gombe chimpanzees and vice versa, and some foods that *are* eaten at both localities are, nevertheless, prepared differently. Driver and Carpenter ants are present at both places: at Gombe the chimpanzees fish frequently for Driver but not Carpenter ants: at Mahale it is the other way round. Gombe chimpanzees use leaf sponges when drinking from water bowls: at Mahale this has not been observed. At Mahale two chimpanzees grooming often show a unique 'hand-clasp' posture never observed at Gombe (McGrew & Tutin 1978). Different traditions have been found in hamadryas baboons and various macaques (see, for example, Kummer 1971) and it seems certain that groups of almost all higher mammals will be found to show intergroup cultural variation. Traditions obviously began with the innovative performance of particular individuals in the past – performances that were subsequently incorporated into the repertoire of a group through processes of social facilitation and observational learning (figure 3.) Only in the Japanese macaques has this process been carefully documented (Kawamura 1954; Itani 1965), first in regard to the unwrapping and eating of candies and subsequently for the dramatic inventions of potato-washing and wheat-cleaning initiated by the gifted young female Imo (and which, as mentioned, led to other sea-related activities). It was young monkeys who were the first to perform the new behaviours, their peers and their mothers who were the first to follow their example. Siblings and other closely associating adults were next on the list. As new infants were born, they acquired the behaviours, in the normal learning process, from their mothers. In the case of the unwrapping and eating of candies the reward was immediate and obvious and it is undoubtedly significant that this behaviour was acquired by all members of the troop, whereas the washing and cleaning techniques were never learnt by some of the adult males.

The rhesus monkeys that were transferred from Santiago to a new island (Morrison & Menzel

1972) rapidly incorporated many new foods into their diet. There was no evidence that these new foods were consistently sampled by any individual monkey or category of monkey: each one tried each new food as he came across it and accepted or rejected it according (presumably) to its taste. When wheat was introduced to one group of Japanese monkeys the whole troop had acquired the new food habit within four hours (Itani 1965). Similarly, the baboons (of all ages and both sexes) at Gombe have always unhesitatingly tried new foods offered them. In chimpanzee society, however, both at Gombe and Mahale, it is only youngsters who experiment with new foods (Nishida *et al.* 1984). Moreover, the conservative attitude of mothers and elder siblings, who usually snatch or flick away a new food item from an infant, makes it unlikely that the infant will repeat the experiment, and even less likely that the food in question will be incorporated into the diet of the community as a whole.

One new behaviour that did spread through the community at Gombe was the use of sticks as levers to try to open banana boxes. Four and a half months after these boxes had been installed three adolescents began, independently, to use sticks to try to prize open the steel lids. Because a box was sometimes opened when a chimpanzee was working at it, the tool use was occasionally rewarded and, over the next year, the habit spread until almost all members of the community, including adult males, were seen using sticks in this way. That many individuals learnt as a result of watching their companions is suggested by the fact that one female was observed to behave thus on her very first visit to camp: before this she had had ample opportunity to watch what was going on from the surrounding vegetation (Goodall 1968).

Most of the innovative performances in the social sphere observed at Gombe were seen in single individuals only, and often only once, as when an adolescent male showed a submissive genital display (facing his aggressor with laterally positioned leg, hip and knee bent and an erection of the penis) exactly comparable to that of the squirrel monkey (Ploog 1967, plate 10.3). A juvenile female, Fifi, also showed a completely new pattern (at least at Gombe, though it was observed in Washoe (Gardner & Gardner 1969)) 'wrist-shaking'. This gesture, a rapid to and fro shaking of the hand, was directed at an adult female in an aggressive context. The following week Fifi repeated the gesture and continued to do so, on and off, for the next ten months, always in the aggressive context. A week after Fifi's first observed wrist-shake the new gesture appeared in a slightly younger female, Gilka, who was Fifi's most frequent playmate. For the next few months Gilka wrist-shook vigorously in many contexts. Then she, too, gradually dropped the gesture from her repertoire (Goodall 1973). Another new pattern was shown by infant male Flint when he inspected the genital area of a female with a small stick, instead of with his finger as is usual. He continued to stick-inspect occasionally over the next few weeks during which time another infant male imitated the behaviour. After this it was not seen again. Köhler (1925) reported a number of novel behaviours which became 'fashions' for a few weeks or months and then vanished from the repertoire of his colony.

Behaviours of this sort, which have no obvious practical reward, are unlikely to spread through a group unless they persist long enough, in one female, for her infant to acquire it during normal learning. The female, Madam Bee, developed a unique grooming pattern, vigorously scratching the skin of her partner before intently examining the skin at that place. The appearance of this behaviour followed the polio epidemic during which Madam Bee lost the use of one arm. At that time her elder daughter was about eight years old, the younger, one and a half years old. Subsequently the younger child was seen using the unique pattern on a number of occasions, but the older daughter was never seen to do so (Goodall 1973).

Twice the characteristic performances of individual males were incorporated into the repertoire of younger males. When Mike was enhancing his display by the use of empty four gallon cans Figan was an adolescent: on several occasions Figan was observed as he 'practised' with an empty can discarded by Mike in the bush. Figan himself, as described, developed a highly effective arboreal display when his companions were still in their nests. Young Goblin, who had associated closely with Figan from early adolescence onwards, and had watched Figan's rise in the hierarchy, subsequently began to perform early morning displays himself when he, at the age of 14, was making a determined bid for power.

In the social sphere, innovations are often designed to get the better of higher ranking companions. And, at least sometimes, they are quickly dropped when counter-strategies are developed. Thus Emil Menzel (1974) describes how a subordinate female, Belle, who had been shown the whereabouts of hidden food, tried, in various and ever more sophisticated ways to withhold this information from the dominant male, Rock (since, if she led him to the place he invariably took all the reward). But Rock quickly learnt to see through her various subterfuges. If she sat on the food, he learnt to search underneath her. When she began sitting half way towards the food, he learnt to follow the direction of her travel until he found the right place. He even learnt to go in the opposite direction when she tried to lead him directly *away* from the food. And, since she would sometimes wait until he was not looking, Rock learned to feign disinterest, but was ready to race after her once she began to head for the goal. This sequence of events is particularly interesting since it shows how one innovation during competitive interactions in the social sphere can stimulate the rival to make a novel response. Such competition between animals as cognitively advanced as chimpanzees can thus lead to an escalation in the development of new social strategies within a group.

Conclusion

More carefully documented 'anecdotal' reports from field studies should yield a wealth of information on innovative performances in the ecological, technical and social spheres. Of the many such behaviours observed, only a few will be passed on to other individuals, and seldom will they spread through the whole troop. However, systematic experimentation (such as the introduction of a variety of carefully designed ecological and technical 'problems' and long-term recording of reactions to them) both in free-living and captive groups would provide a new way of studying the phenomena of innovative behaviours and their transmission through and between social groups.

References

Boesch, C. & Boesch, H. 1981 Sex differences in the use of natural hammers by wild chimpanzees: A preliminary report. *J. Human Evol.* **10**, 585–593.

Boesch, C. & Boesch, H. 1983 Optimisation of nut-cracking with natural hammers by wild chimpanzees. *Behaviour* **34**, 265–286.

Bunnell, B. N., Gore, W. T. & Perkins, M. N. 1980 Performance correlates of social behaviour and organization: social rank and reversal learning in crab-eating macaques (*M. fascicularis*) *Primates* **21**, 376–388.

Bunnell, B. N. & Perkins, M. N. 1980 Performance correlates of social behaviour and organization: social rank and complex problem solving in crab-eating macaques (*M. fascicularis*). *Primates* **21**, 515–523.

Bygott, J. D. 1974 Agonistic behaviour and dominance in wild chimpanzees. Ph.D. Thesis, University of Cambridge.

Chance, M. R. A. & Mead, A. P. 1953 Social behaviour and primate evolution. *Symp. Soc. exp. Biol. VII (Evolution)*, pp. 395–439.
Gardner, R. A. & Gardner, B. T. 1969 Teaching sign language to a chimpanzee. *Science, Wash.* **165**, 664–672.
Goodall, J. van Lawick 1968 The behaviour of free-living chimpanzees of the Gombe Stream Reserve. *Anim. Behav. Monogr.* **1**, 161–311.
Goodall, J. van Lawick 1971 *In the shadow of Man*. London: Collins, and Boston, Houghton-Mifflin.
Goodall, J. 1973 Cultural elements in a chimpanzee community. *Symp. IVth Int. Congr. Primat.* vol. 1: *Precultural primate behaviour* (ed. E. Manzel), pp. 144–184. Basel: Karger.
Goodall, J. 1983 Population dynamics during a 15-year period in one community of free-living chimpanzees in the Gombe National Park, Tanzania. *Z. Tierpsychol.* **61**, 1–60.
Humphrey, N. K. 1976 The social function of intellect. In *Growing points in ethology* (ed. P. P. G. Bateson and R. A. Hinde), pp. 303–317. Cambridge: Cambridge University Press.
Itani, J. 1965 On the acquisition and propagation of a new food habit in the troop of Japanese monkeys at Takasakiyama. In *Japanese monkeys: a collection of translations.* pp. 52–65 (ed. K. Imanishi and S. A. Altmann) Edmonton: University of Alberta Press.
Jolly, A. 1966 Lemur social behaviour and primate intelligence. *Science, Wash.* **153**, 501–506.
Kawamura, S. 1954 A new type of behaviour of the wild Japanese monkeys – an analysis of an animal culture. *Seibutsu Shinka* **2**, 1. (Cited in Itani 1965).
Köhler, W. 1925 *The mentality of Apes*. New York: Harcourt & Brace.
Kummer, H. 1968 *Social organization of Hamadryas Baboons*. Chicago: Aldine
Kummer, H. 1971 *Primate societies: group techniques of ecological adaptation.* Arlington Heights: Harlan Davidson.
Kummer, H. & Kurt, F. 1965 A comparison of social behavior in captive and wild hamadryas baboons. In *The Baboon in medical research* (ed. H. Vagtborg), pp. 1–46. Austin. University of Texas Press.
McGinnis, P. R. 1973 Patterns of sexual behaviour in a community of free-living chimpanzees. Ph.D. Thesis, University of Cambridge.
McGrew, C. & Tutin, C. E. G. 1978 Evidence for a social custom in wild chimpanzees. *Man* (N.S.) **13**, 234–51.
Menzel, E. W. 1964 Patterns of responsiveness in chimpanzees reared through infancy under conditions of environmental restriction. *Psychol. Forsch.* **27**, 337–365.
Menzel, E. W. 1974 A group of young chimpanzees in a one-acre field. In *Behaviour of non-human primates* (ed. A. M. Schrier & F. Srollnitz), vol. 5, pp. 83–153. New York: Academic Press.
Milton, K. 1981 Distribution patterns of tropical plant foods as an evolutionary stimulus to primate mental development. *Am. Anthrop.* **83**, 534–548.
Morrison, J. A. & Menzel, E. W. 1972 Adaptation of a free-ranging rhesus monkey group to division and transplantation. *Wildlife Monogr.* no. 31.
Nishida, T., Wrangham, R. W. & Goodall, J. 1984 Local differences in plant-feeding habits of chimpanzees between the Mahale Mountains and Gombe National Park, Tanzania. (In the press.)
Ploog, D. W. 1967 The behaviour of squirrel monkeys as revealed by sociometry, cioacoustics, and brain stimulation. In *Social communication among primates* (ed. S. A. Altmann), pp. 149–184. Chicago: University of Chicago Press.
Riss, D. & Goodall, J. 1977 The recent rise to the alpha-rank in a population of free-living chimpanzees. *Folia primatol.* **27**: 134–151.
Sigg, H. 1980 *Z. Tierpsychol.* **53**, 265–302.
Schiller, P. H. 1949 Innate motor action as a basis of learning. In *Instinctive behaviour* (ed. C. H. Schiller). New York: International Universities Press.
Schiller, P. H. 1952 Innate constituents of complex responses in primates. *Psychol. Rev.* **59**, 177–191.
Schönholzer, L. 1958 Beobachtungen über das Trinkverhalten von Zootieren. Unpublished doctoral dissertation, University of Zurich.
Strayer, F. F. 1976 Learning and imitation as a function of social status in Macaque Monkeys (*Macaca nemestrina*) *Anim. Behav.* **24**, 835–848.
Tutin, C. E. G. 1979 Mating patterns and reproductive strategies in a community of wild chimpanzees. *Behav. Ecol. Sociobiol.* **6**, 29–38.
de Waal, F. 1982 *Chimpanzee politics: power and sex among apes.* New York, London: Harper Row.
Wrangham, R. W. 1975 The behavioural ecology of chimpanzees in Gombe National Park, Tanzania. Ph.D. thesis, University of Cambridge.

Phil. Trans. R. Soc. Lond. B **308**, 215–216 (1985) [215]
Printed in Great Britain

General discussion

H. J. Jerison (*University of California, Los Angeles* 90024, *U.S.A.*). We have heard surprisingly little during this meeting about evolutionary theory as a source of synthesis for the extensive and somewhat chaotic catalogue of intelligent animal behaviours that have been described. The question raised at the end of the meeting of whether their smaller brains (about one third the size of human brains) limit the capacities of chimpanzees to learn human language is clarified if treated as an evolutionary question. It is on grades (not 'clades') of phenotypic evolution.

Phenotypically, language may be defined by its status in normal human adults. The phenotype has its ontogeny, of course, as well as a variance and 'mean value'. (Pathologies that generate outliers are excluded from the analysis.) Brain size is associated with language in living humans primarily because so large a fraction of the brain (perhaps 50 %) has been associated with language and language-like functions, which include functions of the 'minor' hemisphere. Evidence on the evolution of brain size that is legitimately applied to issues about the evolution of language must be based on 'mean values' of brain size. Similarly, if one considers brain size in chimpanzees as relevant for such evidence the comparison must be between mean values for chimpanzees and mean values in humans. It is appropriate to note, for such an argument, that the coefficient of variation for brain size is about 10 % for both humans and chimpanzee adults, so variability is not at issue.

In this perspective, it is unlikely that the impressive results reported on chimpanzee 'language' refer to the same adaptation in chimpanzees as in humans. Rather they are better interpreted as showing that we have badly underestimated the cognitive capacities of chimpanzees, and that teaching them to communicate with a system of signs derived from human language has enabled us to form a better idea of their cognitive capacities. This entry into the 'mind of the chimpanzee' (opening, also, for some cetaceans according to Herman's recent work) is an important accomplishment by comparative psychologists, not to be muddied by raising the issue of language.

L. Weiskrantz, F.R.S. One of the points to have emerged from this meeting, I believe, in exposing the current 'state of the art', is that the issues involved in trying to relate the cognitive communicative capacities of the chimpanzee to the achievement of 'language' are not only still open, but are more complex than was imagined 20 years ago. One question, for example, is whether 'intentionality' is a necessary criterion for attributing *linguistic* communication and, if so (which is by no means agreed), how intentionality can be recognized and identified. It seems to me that the behavioural criteria are likely to be ostensive rather than logical, with the attendant difficulties in resolving disagreements about judgments at the empirical and observational level. Another question concerns a stimulus that has become associated, by pairing, with reward or punishment: when does it qualify as a meaningful symbol and, especially if it is generated by the animal itself (as in making an A.S.L. sign), when does it become a word? Incidentally, given the difficulty that human subjects have in mastering lists of 'paired-associates' and the not unformidable task it is for an animal to acquire lists of neutral

S-S associations, it is somewhat paradoxical to see the phrase 'mere paired-associate learning' being used pejoratively! That aside, the empirical evidence itself as regards the capacity of animals to manipulate strings of tokens or A.S.L. signs is far from complete and, as we have seen, is still controversial, although I believe the issues have become clarified as a result of the meeting.

These issues cannot be settled by making assumptions about or appeals to brain size *per se*, because it will be necessary to say what it is, logically and empirically, that allows one to judge when a brain has reached a 'threshold' size in evolution to grant it the capacity for language (leaving aside the questions that arise from considerations about continuity in evolution). Nor would there be agreement that size *per se* is the critical factor: there are those who argue that there is a particular, possibly unique, organization of the human brain that endows it with linguistic capacity even when the brain is undersized in other respects through constitutional disorders or genetic mistakes. Nor is it easy to estimate the fraction of the human brain associated with language – this can hardly be 50% given that the demonstrable 'language' regions constitute much less than the whole of one (the dominant) hemisphere – but stretching the category to include 'language-*like* functions' begs the question of just what cognitive skills are 'language-like' and how they might differ from 'language', especially as Professor Jerison grants that the cognitive capacities of the chimpanzee allow it to communicate with a system of signs. And so we return to the very issues that are as yet unresolved.

But all will agree with Professor Jerison's remark that the work of contemporary comparative psychologists, much of it reviewed here, has provided an important entry into the 'mind of the chimpanzee' (and not only the chimpanzee). As always, the end of one meeting makes one regret that it could not be followed immediately by a second that would concentrate on those issues revealed but left unresolved by the first, and might also allow the evidence to be extended more widely to include not only the Cetaceans but also the invertebrates such as *Aplysia*, on which so much exciting basic neurobiological work on some forms of learning and memory is now being done.

AUTHOR INDEX

(*Page numbers in italics refer to citations in reference lists*)

SUBJECT INDEX